Dialogues and Games of Logic

Volume 4

Hintikka's Take on Realism and the Constructivist Challenge

Dialogues and Games of Logic Series Editors
Shahid Rahman   shahid.rahman@univ-lille3.fr
Nicolas Clerbout
Matthieu Fontaine

Hintikka's Take on Realism and the Constructivist Challenge

Radmila Jovanović

ISBN 978-1-84890-194-0

College Publications
Scientific Director: Dov Gabbay
Managing Director: Jane Spurr
Department of Informatics
King's College London, Strand, London WC2R 2LS, UK

www.collegepublications.co.uk

Original cover design by Laraine Welch
Printed by Lightning Source, Milton Keynes, UK

Table of contents

Acknowledgements

The present book is based on my PhD Thesis, which was defended publicly on the 9th of February, 2015 at the University of Lille 3.

The writing of a thesis is a unique adventure in life. On the one hand, it requires many lonely hours of quiet work in isolation, but on the other hand—fortunately—it demands a lot of interactions and exchanges with colleagues working in the same field. I was so fortunate to work in very friendly and inspiring environments. I want to mention at least some of those wonderful people who contributed to my work and who helped me to find and enjoy my path.

This thesis was realized following the co-tutelle program between the University of Belgrade and Charles de Gaulle University, Lille 3, under the supervision of two great professors, Dr Milos Arsenijević and Dr Shahid Rahman. Their influence on my work and on my professional development are difficult to overestimate.

I met professor Arsenijević in Belgrade during my graduate studies. He awoke my love for philosophy and critical thinking and his influence was incredibly important. I am so grateful for his support and all the opportunities he gave me to collaborate with him.

During my studies my dear friend Ivan Vuković (associated profesor at the Faculty of Philosophy in Belgrade) introduced me to the network OFFRES (Organisation Francophone pour la Formation et la Recherche Européennes en Sciences Humaines), where, among other dear friends and colleagues, I met professor dr Patrice Canivez, a very kind person who introduced me to Prof. Dr Shahid Rahman. I was so fortunate to obtain two scholarships from the French Government, one for a master's degree and another for a doctorate, which allowed me to go to Lille and to join the incredible group of logicians that had been assembled by professor Rahman. My adventure really began there. I am infinitely grateful to Professor Rahman for giving me the once-in-a-lifetime opportunity to work under his guidance. Not only is one of the foremost logicians and great philosopher, he is also a uniquely kind and inspiring person who managed to bring together a very fruitful group of "dialogicians" and create a friendly and supportive atmosphere (maybe the only atmosphere in which the sciences can really flourish).

My work rests on recent work in logic done by professor dr Rahman and dr Nicolas Clerbout who made a connection between Constructive Type Theory and Dialogical logic. Both their work and mine were inspired by a visit from professor dr Göran Sundholm to Lille, where he gave a series of lectures. My work was in great part motivated by these lectures and discussions I had with him in person and by mail.

I was also very privileged to have a highly distinguished jury for the defence of my thesis: beside my supervisors, there were Dr Andrej Jandrić, Dr Mark Van Atten, Dr Michel Bourdeau, and Dr Gerhard Heinzmann. I want to thank them very much for their feedback, remarks, and questions, for which I am very grateful.

To come back to the begging, as the writing of a thesis is not always an easy task and engages us on so many levels, it would be difficult to accomplish it without the support of close friends and family. I dedicate this book, as well as the thesis, to my parents, who made everything possible. I thank them, my sister, and my wider family. I have also been blessed with most wonderful and caring friends, and I thank them for all their love and encouragement. In the end, I owe special thanks to my husband Michal, for love is the biggest inspiration of all.

Introduction

Mathematical game theory was first introduced by John von Neumann and Oskar Morgenstern in 1944.[1] During the second half of the twentieth century the link was made between games and logic and, since then, game theory has proven to be the most fruitful framework for different areas of logic and mathematics, as well as for the theory of argumentation. A specific kind of game—two person, win or lose—found their application in providing a new kind of rule-based semantics and since then have served different purposes in logic—among others, in providing a definition of truth or in providing an account of formal proofs.

Game-theoretically oriented semantics, which I am interested in here, provide an alternative to traditional Tarski-style semantics, implementing Wittgenstein's idea of *the meaning as use*. The basic idea is that the meaning is obtained in a game between two players, one trying to defend and the other trying to falsify the expression at stake. The notion of truth, or that of validity, is based on the existence of a *winning strategy* of the initial verifier in a game. The direction is the opposite of that in Tarski-style semantics: the game starts with the entire expression and runs until its component parts are reached. In this book I will be interested in two different game theoretical traditions: Game Theoretical Semantics, developed by Jaako Hintikka and Gabriel Sandu, and Dialogical logic, first introduced by Paul Lorenzen and Kuno Lorenz and further developed by Shahid Rahman and his associates.

By 1960 *Dialogical logic* was first introduced by Lorenzen and Lorenz and was concieved as a solution to some of the problems that emerged in Lorenzen's *Operative Logik* (1955).[2] Proof theoretical approaches led to the so-called *epistemic turn*, which came to be linked with the notion of games and provided the dynamic features of traditional dialectical reasoning. Inspired by Wittgenstein's *meaning as use*, the basic idea of the dialogical approach to logic is that the meaning of the logical constants is given by the norms or rules for their use. The approach provides an alternative to both model-theoretic and proof-theoretic semantics. Originally, dialogues were developed in the context of constructive mathematics and logic but since then they have served the comparison and combination of different logical systems.

---

[1] (von Neumann, Morgenstern, 1944)

[2] The main original papers are collected in Lorenzen and Lorenz (1978). Other papers have been collected in Lorenz (2010a,b).

In 1968 Jaakko Hintikka combined the model-theoretical and the game-based tradition to create his Game Theoretical Semantics[3] (GTS) that, like in the dialogical framework, grounds the concepts of truth or validity on the game - theoretic concepts, such as the existence of a winning strategy for a player, but differently to the dialogical framework it is related to the notion of model. Hintikka claims that his semantic games are exact codifications of language games in Wittgenstein's sense, at least if one accepts that the activities associated with quantifiers are *looking for* and *finding* appropriate individuals in the model. The inspiration came partially from Peirce, who had already suggested in his second Cambridge Conference lecture of 1898[4] that the difference between *every* and *some* can be explained in terms of who chooses an object. In the late fifties Henkin proposed that we could understand in game terms some sentences that had not had satisfactory treatment in Tarski's semantics, such as those containing an infinite string of quantifiers.[5] Henkin came up with the idea of a game for such a string of quantifiers that would take place between two players who choose objects to replace bound variables, and he suggested that the truth conditions of the sentence could then be expressed using its Skolem version. Hintikka generalized this idea to come up with a Game Theoretical Semantics for classical first-order logic, which has been proven to be equivalent to Tarski-style semantics, given that the axiom of choice holds. That is, linking a first-order sentence with its truth conditions, namely its Skolem form, yields a version of the axiom of choice, which proves, according to Hintikka, that the axiom of choice is an unproblematic first-order principle. Games for classical first-order logic are deterministic games of finite length,[6] and they are games of *perfect information*, where the players are aware of all of the previous moves in a game. In 1989 Hintikka and Sandu came up with games with *imperfect information*, where a restriction can be made on the information available to a player—so that the player sometimes has to play without knowing (or by forgetting) previous moves made by the opponent. Those games yielded *Independence friendly first-order logic* (*IF logic*), exceeding the expressive power of classical first-order logic. The great expressive power is due to the possibility of expressing, in IF language, more dependency patterns between quantifiers and other logical constants than in standard logic. It is expressive enough to enable formulating linearly, and at the first-order level, sentences containing branching quantification. Because of this characteristic, Hintikka claims that IF logic is most suitable for at least two main purposes:

---

[3] Hintikka (1962, 1973, 1996a), Hintikka and Sandu (1997). See also Hintikka (1999) and in particular Hintikka et al. (1999). Rahman and Tulenheimo (2006) studied the relation between dialogical logic and GTS.

[4] In Peirce (1898).

[5] In Henkin (1961). See also Hodges (2013).

[6] Which is proven by the Gale-Stewart theorem.

1) to be the logic of the first-order fragment of natural language; and
2) to be the medium for the foundation of mathematics.

This book aims to explore the above uses of IF logic. As for the first point, Hintikka's argument is that, since it is more expressive than standard classical logic, in IF language some natural language sentences (that are perfectly first-order) can be expressed that otherwise could not be expressed in classical first-order logic. Though Game Theory Semantics provides a satisfactory account of some language phenomena, such as anaphora, to which it is difficult to give a suitable interpretation otherwise, Hintikka claims that at least some sentences containing anaphoric expressions can *only* be treated in a satisfactory manner using the Independence Friendly logic formalization.

In relation to the second use of IF logic, that of being suitable for mathematical reasoning, Hintikka develops this suggestion in detail in his work *The Principles of Mathematics Revisited* and in many papers following this major work. He argues there that IF logic is a superior alternative to set theory (usually formalized with classical first-order logic) and that, linked with game-theoretical semantics, it provides a realist foundation of mathematics, embracing the greater part of classical mathematics. It is, in his opinion, *a dream of logicists come true*, a framework allowing the conceptual reduction of mathematic to logic, as the very title of his main work suggests, as a clear allusion to Russell and Whitehead's work that set a landmark in the field. Moreover, Hintikka insists that in GTS, as mentioned above, Zermelo's axiom of choice is justified on the first-order level in a way that should be perfectly acceptable for constructivists.

However, IF logic has some serious limitations, which I will discuss in this thesis. One of the biggest is the impossibility of full axiomatisation. The lack of a complete proof system means that IF logic is inevitably linked—if not reducible—to to model theory. Although Hintikka doesn't recognise this as a problem, it seems to clash with his own claim that *IF-logic is the logicist's dream come true*.

Moreover, according to Ranta and Sundholm, game theoretical semantics allows a constructivist reading as well, but Hintikka is not in favour of the constructivist approach. One of the principle reasons for his rejection of constructivist, proof-conditional theory of meaning is Hintikka's claim that in this framework one cannot give a suitable account of branching quantifiers. I will prove him wrong in regard to this point. But I will also raise a question about the possibility of a constructivist reading of his Game Theoretical Semantics in general. I will suggest that this reading would be difficult, if not impossible, to endorse.

In this work I will present, analyse, and criticize the above-mentioned uses that Hintikka makes of IF logic. Before I elaborate on the main thesis, let me briefly cover the dialogical tradition I have previously mentioned and its recent developments, which will be my main tool for refuting Hintikka's theory. The formal tool I will use is a dialogical approach to *Constructive Type Theory*, introduced by Per Martin-Löf.

In opposition to the model-theoretic tradition base of Frege's and Tarski's work, where set theory was linked with the classical first-order logic, in the second half of the twentieth century a constructivist approach to mathematics and logic was born in the context of Brower's intuitionistic foundations of mathematics. In this stream an epistemic approach to meaning was brought to the fore. Emphasis was placed on the *knowledge* and on the conditions for a proposition to be *asserted* rather than on the *truth* of a proposition, as in the old tradition. This way of thinking demanded a *proof theory* in opposition to a model-theoretic semantics. In this tradition Martin-Löf developed Constructive Type Theory (CTT), where the principle *"propositions-as-types"* has a central place. According to this principle, propositions are identified with sets, types, or sets of their proof-objects. CTT has intuitionistic logic at its core and it is fully predicative. Some of the important features of CTT are:

- In CTT a clear distinction is made between judgments (or assertions) and propositions expressed by those judgments.

- Classical first-order logic can be developed in the CTT framework (see Ranta, 1994: Chapter 2).

- As opposed to the model-theoretic approach to meaning, where elements of language are linked to the world (or interpreted) via some meta-level means, in CTT judgments and proof objects are embedded in the object-level, and they account, by means of inferential rules, for the meaning of every expression, such a fully interpreted language results. This carries out Frege's idea of *Begriffsschrift*—which should provide a full and explicit formalization of mathematics.

Recently, a link has been established between CTT and dialogical logic (see Rahman, Clerbout, 2013, 2014), thus putting CTT in a game-theoretic framework. It seems to be a powerful formal tool, although work on it has just started. As mentioned above, the idea behind CTT is to render meta-logical features explicit at the object-language level. This is actually very close to

Wittgenstein's arguments against formal semantics in the style of Tarski,[7] and is one of Wittegenstein's tenets that Hintikka (and other model-theoreticans) rejects. Moreover, the aim of rendering the rules of meaning-explanation explicit is inherent to the dialogical approach, where meaning is constituted within certain interactions between players. In such a context it is natural to request that those interactions be rendered explicit in object-language. It should also be pointed out that, though dialogical logic shares some basic elements with GTS, there are at least two main distinctions, namely: (1) dialogical logic does not assume an underlying model—as GTS does—and this links dialogical logic to constructivist approaches in general; (2) it distinguishes between the *local meaning* obtained at the play-level, and the *global meaning* obtained at the level of strategies. As it happens, the combination of CTT with dialogical logic proves Hintikka wrong in his claim that constructivist approaches can provide neither an account of branching quantifiers in the first-order level nor the outside-inside reading of expressions containing such kind of quantification.

Finally, let me list the main theses to be developed in the present book:

1) I will present and analyse two main uses Hintikka makes of IF logic and GTS: one as the medium for mathematical reasoning and the other as the means of formalisation and analysis of natural language. The properties of IF logic will be discussed, as well as the advantages of this approach such as the possibility of taking account of (in)dependency relations among variables; GTS accounts for two different notions of scope of quantifiers; the "outside–in" direction in approaching the meaning, which turns out to be advantageous over the traditional "inside-out" approach; the usefulness of game-theoretic reasoning in mathematics; the expressiveness of IF language, which allows formulating branching quantifiers on the first-order level, as well as defining the truth predicate in the language itself. I will defend Hintikka's stance on the first-order character of IF logic against some criticisms of this point.

2) The weak points will also be discussed: first and foremost, the lack of a full axiomatization for IF logic and second, the problem of signalling, a problematic phenomenon related to the possibility of imperfect information in a game.

3) I will turn to another game-theoretically oriented semantics, that of Dialogical Logic linked with Constructive Type Theory, in which

---

[7] For a discussion on this see Rahman, Clerbout, Jovanović, forthcoming.

dependency relations can be accounted for, but without using more means than constructive logic and the dialogical approach to meaning have to offer. I will use this framework first to analyse Hintikka's take on the axiom of choice, and second to analyse the GTS account of anaphora.

4) I will compare and contrast Martin-Löf's analysis of the axiom of choice with Hintikka's stance towards this axiom. Hintikka claims that his game theoretical semantics (GTS) for Independence Friendly Logic justifies Zermelo's axiom of choice in a first-order way perfectly acceptable for constructivists. Martin-Löf's results, however, show that Hintikka's preferred version of the axiom of choice is indeed acceptable for constructivists and that its meaning does not involve higher-order logic if it is based on an intensional take on functions. But the heart of the classical understanding of Zermelo's axiom is extensionality, and this is the real reason behind the constructivist rejection of it. Furthermore, I will show that the dependence and independence features that motivate IF logic can be formulated within the frame of Constructive Type Theory (CTT), without paying the price of having a system that is neither axiomatizable nor has an underlying theory of inference.

5) I will analyse the GTS account of anaphora. I will argue that, though the GTS approach has considerable advantages over other existing theories in dealing with anaphora, the extension of the dialogical framework contains both the contentual (first-order) features of CTT and the interactive aspects of GTS. In the dialogical approach, choice dependencies are expressed at the object-language level and they can be expressed in such a way that: 1. no means are needed other than those that constructive (or even classical first-order) logic has to offer; 2. there is no possibility of some non-desirable phenomenon, such as *signalling*.

Consequently, the aim of this book will be to show how a satisfactory treatment can be made in CTT dialogues for both anaphora and branching quantifiers on the first-order level, and how relations of (in)dependencies between quantifiers can be accounted for, but without sacrificing the inferential role of logic. It will be equally clear that neither branching quantifiers nor the anaphora push us towards a model theory.
In the the following chapter I provide a detailed overview of this work.

An overview of this book

In part I of this work I will present *Independence Friendly logic (IF)* and two of the most cherished uses Hintikka makes of it: one in the foundation of mathematics and the other in the formalization of a natural language in general and of anaphoric expressions in particular.

Firstly, I introduce the IF language, starting with classical first-order language and enlarging it with a new syntactical device that makes a crucial change: a slash "/" indicating the relation of independence of a quantifier or a connective in relation to some quantifier within whose scope it lies.

The semantics of IF language is given in section I.3., where I present *Game Theoretical Semantics* (GTS). The traditional Tarski-style, compositional semantic is inapplicable in IF logic, where formal barriers, involving the slash sign, prevent it from conforming to such a compositional framework. A satisfactory, non-compositional, model-theoretic semantics is given in the game-theoretical framework: the satisfaction of an IF sentence is checked in a model through a game between two players, one trying to verify and the other to falsify the sentence. The game rules are presented and explained in this section. In general, I will follow Hintikka's own presentation, but I will introduce some changes, following other works dealing with IF logic, with the appropriate theoretical justifications.

In section I.4. I will discuss in more detail the games in question, presenting them in a general game-theoretic framework. For this I will use two main sources: Mann, Sandu, and Sevenster's (2011) book, and a thesis by Dechesne (2005). Games related to IF logic are classified in the literature as *extensive games with imperfect information*. I will explain in this section two crucial notions related to IF logic and GTS: that of restriction in the information available to a player and that of a strategy.

In section I.5. I will discuss some of the properties of IF logic. This section is thus divided into fifteen sub-sections, in each of which I try to shed light on different features of IF logic and GTS that will be of relevance for later discussions in this work (the analysis of IF logic could certainly be enlarged upon, but I will limit the discussion to what is relevant for my purposes). I will start with the notion of *strategy* in a game, which leads to the analysis of the *axiom of choice* included in the very definition of truth in GTS. That is, the truth of a given sentence is defined as the existence of the winning strategy of the initial verifier in a game for that sentence, where strategies are expressed by Skolem functions. Linking the sentence with its truth conditions, that is, with the assertion that such a choice function exists, yields a variant of the axiom of choice. One of Hintikka's favorite examples of the supremacy of GTS is what he considers to be a justification of Zermelo's axiom of choice in a way that is acceptable for a constructivist. Hintikka's take on the axiom of choice will be one of my main targets in this thesis.

I will then move on to the Skolemization procedure, related to the truth conditions of a sentence and the procedure of Kreiselization, related to its falsity conditions. I will follow this with a discussion of the failure of the law of the excluded middle in IF logic. I will explain the reasons for the failure of this classical law and I will compare it to the case of intuitionistic logic, where this law happens not to be valid for entirely different reasons.

I will then make some remarks on the expressive power of IF logic, which significantly exceeds that of classical first-order logic, even though, according to Hintikka, no commitment to higher-order entities is involved in it. This is the reason why Hintikka claims that IF logic is more convenient for the formalization of the first-order fragment of natural language. IF logic is expressive enough to embrace the branching quantification on the first-order level, thus making it the best candidate both for foundation of mathematics and for the formalization of natural language: all important mathematical concepts can be expressed in IF logic, as well as some natural language sentences involving *anaphoric expressions* that could not be expressed otherwise without involving a higher-order quantification. Both proposed uses of IF logic will be discussed in this book.

I will continue by examining the failure of axiomatizability and the impossibility of giving a compositional interpretation of IF logic. Hintikka does not consider either of these to be handicaps (the lack of compositionality is rather taken to be an advantage). His justification for the first is made from a model-theoretic perspective, while the lack of compositionality is justified by the desirability of the meaning's depending on the context in which the language is used. I will make some remarks about this.

In the next sub-section I discuss the possibility of formulating a truth predicate for IF language in the language itself—one of the biggest advantages of IF language, according to Hintikka. That is, as per Tarski's famous result from 1936, the truth predicate cannot be defined for a first-order language within that same language because adding a truth predicate in a first-order language would produce paradoxes. The possibility of formulating a truth predicate for IF first-order language is another important argument Hintikka uses in favour of the application of this logic in the foundation of mathematics.

I will consider an extension of IF logic, where another kind of negation is introduced—which ought to further enlarge the expressivity of IF logic and thus make it completely suitable for mathematical purposes.

Furthermore, some other properties will be proven to apply to IF logic, such as the law of double negation, de Morgan's laws, compactness, the Löwenheim-Skolem theorem, and the separation theorem. In relation to the Löwenheim-Skolem theorem, I will briefly discuss the so-called Skolem's paradox. Even though Hintikka recognizes, as do most mathematicians and philosophers, that this is not the real paradox because it does not produce any real contradiction in mathematical theories, he creates an argument that uses it against the usual first-order formalization of set theory. Given the fact that the Löwenheim-Skolem theorem also holds for IF logic, and given that Hintikka claims that it is a genuine first-order logic, it is pertinent to look at Hintikka's solution to Skolem's paradox. Finally, I will make a short remark on the notion of equivalence in IF logic, which is specific, given that the law of the excluded middle does not hold.

Some problems with IF logic will be discussed in section 1.6. The first is a phenomenon appearing in some games for IF sentences called *signalling*. In many cases this phenomenon is undesirable and Hintikka tries to prevent it using certain formal restrictions. However, it seems that the problem is not so easily ruled out because, as it happens, signalling is actually needed in some other cases—for without it, the truth value of some classical first-order sentences in GTS would differ to those under the Tarskian interpretation, which is of course unacceptable.

Another problem concerns the order of IF language. Hintikka's claim that we are dealing here with plain first-order logic has been disputed by some authors. Feferman (2006) and Väänänen (2001) raised the question of whether IF logic is really first-order logic, or whether it is in fact second-order logic in disguise. Tulenheimo (2009) provides some elements to defend Hintikka's view. Curiously, Sundholm (2013) shows that those dependencies and independencies that motivate Hintikka's introduction of IF logic can be formulated in CTT-first-order settings, thus defending Hintikka's ground. With regard to this issue, I will also consider some earlier arguments concerning the order of branching quantifiers: that of Patton, Quine, and Hand.

The last part of the first section serves to connect the dots presented previously in order to offer a clearer picture of Hintikka's take on the role of IF logic in general and of the project of the foundation of mathematics in particular. Hintikka's take on the foundation of mathematics will be analysed: both the specific realist position and the logicist one, although in a modified and less ambitious version: as a conceptual reduction (not a translation) of mathematics to

IF logic. I will also discuss Hintikka's stance towards constructivism and his view of the ontological status of mathematical objects. With this section the first chapter will be complete. In what follows some of Hintikka's crucial points will be contested.

In the second chapter I will present Martin-Löf's Constructive Type Theory in the dialogical framework. The aim of this chapter is to provide some convenient formal tools, which I shall use in the following chapters to question some of Hintikka's points.

In the first part some features of Constructive Type Theory (CTT) will be briefly commented upon. In the second part I will give an outline of the dialogical approach to logic and I will make a few remarks on the relation between GTS and dialogues. As two game-theoretical frameworks, they apparently have a lot in common, but there are some important differences to be stressed as well. A dialogical presentation of classical first-order logic can be found in the Appendix.

In section II.3. I will present recent developments in dialogical logic, where a link is made between dialogical logic and CTT from Rahman, Clerbout (2013, 2014) and Rahman, Clebourn, and Jovanović (forthcoming). Accordingly, in this section I will present the rules for dialogical CTT-games: these include rules for the formation of propositions (which are necessary to satisfy the request of a fully interpreted language), particle rules (that yield, together with the formation rules, local meaning), and structural rules (which determine a general course of a dialogical game—where the global meaning is achieved).

In chapter III of this book I will compare and contrast Martin-Löf's analysis of the axiom of choice with Hintikka's standing on this axiom. Hintikka claims that his game theoretical semantics for IF logic justifies Zermelo's axiom of choice in a first-order way perfectly acceptable to constructivists. In fact, Martin-Löf's results lead to several very important considerations: that there is a reading of Hintikka's preferred version of the axiom of choice which is indeed acceptable for the constructivists and its meaning does not involve higher-order logic, but this reading is based on an intensional take on functions. The problem is that extensionality is necessary for the classical understanding of Zermelo's axiom and this is the real reason why constructivists reject it. More generally, it is possible to formulate the dependence and independence features that motivate IF logic within the frame of CTT without being forced to accept the unpleasant consequence of having a system that is neither axiomatizable nor has an underlying theory of inference.

Thus, my aim is first to make a recapitulation of Hintikka's take on the axiom of choice and then to contrast it with Martin-Löf's (2006) point: that there are indeed some versions of the axiom of choice that are perfectly acceptable for a constructivist, specifically one where the choice function is defined *intensionally*. In order to see this, the axiom must be formulated within the frame of a CTT-setting. Indeed such a setting allows for comparing the extensional and the intensional formulation of the axiom. It is in fact the extensional version that implies the law of the excluded middle, whereas the intensional version is constructive.

In section III.3. I will present Rahman and Clerbout's (2014) dialogical proof of the constructive formulation of axiom of choice. With this they proved Martin-Löf's point in a game-theoretical framework. This result leads us to conclude that the CTT approach to meaning in general, and to the axiom of choice in particular, is very natural to game theoretical approaches where (standard) meta-logical features are explicitly displayed at the object-language level. Thus, in some way, this vindicates, although in a different manner, Hintikka's plea for the fruitfulness of a game-theoretical semantics in the context of the foundation of mathematics. I conclude that the proof of the axiom of choice is constructive but its game theoretical interpretation is antirealist after all.

In Chapter IV I examine Hintikka and his associates' work on anaphora, based on GTS. In this section I compare the GTS approach to the problem of anaphora with a solution provided by the dialogical approach to CTT. I believe that the GTS approach has considerable advantages over other existing theories in dealing with anaphora. However, my aim is to show that the extension of the dialogical framework discussed in the preceding sections contains both the contentual (first-order) features of CTT and the interactive aspects of GTS. The dialogical approach provides a first-order solution that does not require any devices other than those of constructive or even classical first-order logic, when formulated within a suitably adapted CTT-frame. At the end of the chapter I show how the more difficult examples involving branching quantifiers can also be handled in the dialogical framework for CTT in a satisfactory manner.

To finish, I will summarize the conclusions of this work and I will make some further remarks about possible future research.

PART I

Independence- friendly logic

Independence friendly first-order logic (a.k.a. *IF logic*) was introduced by Jaakko Hintikka and Gabriel Sandu in their article 'Informational Independence as a Semantical Phenomenon' (1989). Early works on this form of logic are (Hintikka, 1991) and (Sandu,1991). IF logic is an extension of classical first-order logic, where a new symbol is introduced, '/' (slash, the independence indicator), which serves to express at the object-language level the relation of independence of a quantifier or a connective in relation to a quantifier within whose scopes it lies.

I. 1.1. Classical first-order logic

Let me start by defining classical first-order language in the usual way with the following sets:

$S_1 = \{x, y, z... \}$ — a countable set of variables.

$S_2 = \{\neg, \wedge, \vee, \exists, \forall \}$ — a set of logical symbols for negation, conjunction, disjunction, existential and universal quantifier, respectively.

$S_3 = \{ R_n: n$ is an element of the set of natural numbers N $\}$ — the set of relational symbols. Every such relational symbol R has a natural number n that indicates its arity. Unary relational symbols are *predicates.*

$S_4 = \{ f_m :$ m is an element of the set of natural numbers N $\}$ — the set of functional symbols. Every such functional symbol f has a natural number m that indicates its arity. Nullary function symbols are *constant symbols,* which will be defined in a separate set.

$S_5 = \{c_1, c_2 ... \}$ — the set of constant symbols.

$S_6 = \{ ( ) , = \}$ — the set of additional symbols.

I will introduce the set σ as a *signature* that consists of a union of S_3, S_4, and S_5, or the set of all non-logical symbols. Let this be the language L. Now I can define L-terms.

Definition 1:

> a) Every variable and every constant symbol belonging to σ is an L-term;
>
> b) if f is an n-ary functional symbol from σ and $(t_1 ... t_n)$ are L-terms, then $f(t_1 ... t_n)$ is an L-term. I will simply write *terms* from now on.

By combining terms with relational symbols we arrive at *atomic L-formulas*. By combining formulas with logical connectives and quantifiers we get *compound L-formulas*. For the sake of simplicity I will refer to these as *formulas*.

Definition 2:

> If R is n-ary relational symbol from σ and $(t_1 ... t_n)$ are L-terms then $R(t_1 ... t_n)$ is an atomic formula.

Formulas are defined as follows:

Definition 3:

> a) Every atomic formula is a formula.
>
> b) If t_1 and t_2 are terms then $(t_1 = t_2)$ is an atomic formula.
>
> c) If φ is a formula then ⌣φ is a formula.
>
> d) If φ and ψ are formulas then $(φ ∧ ψ)$ and $(φ ∨ ψ)$ are formulas.
>
> e) If φ is a formula then $(∃xφ)$ and $(∀xφ)$ are formulas.
>
> f) Nothing else is a formula.

When they are not necessary I will leave the brackets out. I still need a definition of a *free* and *bound* occurrence of a variable.

Definition 4:

> The occurrence of a variable x is *bound* if it lies within the scope of a quantifier ∃x or ∀x; otherwise it is free. Sets of all free and of all bound variables of a formula φ we call Free (φ) and Bound (φ), respectively.

In atomic formulas all variables occur free. For compound formulas:

Definition 5:

Free $(\smile\varphi)$ = Free (φ);

Free $(\varphi \wedge \psi)$ = Free (φ) U Free (ψ);

Free $(\varphi \vee \psi)$ = Free (φ) U Free (ψ);

Free $(\exists x \varphi)$ = Free (φ)-$\{x\}$;

Free $(\forall x \varphi)$ = Free (φ)-$\{x\}$.

A formula with no free occurrence of variables is a *sentence*.

I.1.2. Models or structures

A structure or a model of language L is a non-empty set of objects with relations and operations that interprets the symbols in the set σ.

Definition 6:

A *structure* or a *model* of vocabulary L is an object $\mathbb{M} = (M; R^M..., f^M ... c^M ...)$ where M is a *universe* of $\mathbb{M}$ which is non-empty;

R^M is an n-ary relation in the universe, which is an interpretation of an n-ary relational symbol R;

f^M is an n-ary function on M,

$f^M : M^n \to M$, that is an interpretation of functional symbol f;

c^M is an element of the universe M that is an interpretation of a constant symbol c.

Definition 7:

A function from set of individual variables to the universe M for a given structure $\mathbb{M}$ is a *valuation v*. If *a* is an element of M and *v* is a valuation then $v(x_i / a)$ is defined as: $v(x_i / a)(x_j) = v(x_j)$ if $i \neq j$; otherwise: $v(x_i / a)(x_j) = a$ if $i = j$.

I will allow for a valuation to also include a mapping from terms to individuals of the universe, so that we have: $v (c) = c^M$ and $v(f(t_{1....} t_n) = f^M (v(t_1)...v(t_n))$.

Definition 8:

> A model is *suitable* for a formula φ if it has interpretations of all relational and functional symbols in that formula.

Now I can state when an atomic formula is satisfied in a model.

Definition 9:

$$\mathbb{M}, v \models (t_1 = t_2) \text{ iff } v (t_1) = (t_2) ;$$

$$\mathbb{M}, v \models R(t_1...t_n) \text{ iff } (v (t_1)... v (t_1)) \in R^M.$$

I.2. Independence friendly logic-syntax

As previously mentioned, syntactically, IF logic has all the elements of classical first-order logic, with the addition of one new symbol. The only new symbol that appears in IF logic is the slash (/), which serves to express the independence of a quantifier or a connective in relation to some quantifier within whose scopes it lies. For example:

1)$\forall x (\exists y/\forall x) S (x,y); \exists x (\forall y/ \exists x) S (x,y)$

2)$\forall x (Sx (v/\forall x) Rx); \exists x (Sx (\wedge/\forall x) Rx)$

I will explain the meaning of the slash sign in detail in the section regarding semantics but let me just touch on it briefly here. In classical logic, the only means of expressing relations between quantifiers are brackets and the nesting of scopes. For Hintikka this is an important limitation, especially when it comes to mathematics. In some sense, the most interesting part of logic is precisely the understanding of relations of dependence and independence among quantifiers (and other logical operators) and the logic that allows expressing these relations has an undoubtable advantage. Expressing relations of dependency entails a completely different understanding of the nature of quantifiers. According to Hintikka (2008:

1), a simplified view on quantifiers is a mistake rooted in Frege's work. Frege interpreted the existential quantifier $\exists x$ in the expression $\exists x S(x)$ as the second order predicate that indicates whether the first-order predicate S is empty or not. The expression is then satisfied if there is an individual a to which the predicate S applies such that $S(a)$ is true. However, this is not the only role quantifiers might have: their other, equally important, role is to express dependencies between variables. When a quantifier is applied to a more than one-place predicate, as in the example $\forall x \exists y S(x,y)$, we face a dilemma: to find a value of the variable y, do we look for it knowing or not knowing the value of x? According to Hintikka (2008: 3), it was Pierce who actually recognised this important semantic aspect manifested in the ordering of quantifiers, which led him to the concept of a semantic game for quantifiers played by two players. This was one of the inspirations for Hintikka's *Game Theoretical Semantics*.[8] IF logic takes account of quantifiers' dependency relations and so allows for freeing the expressivity of a first-order language.

Now let us return to the syntax of IF logic. We need to make some changes in definition 3 for IF formulas:

Definition 3*:

    a) Every atomic formula is a formula.

    b) If t_1 and t_2 are terms then $(t_1 = t_2)$ is an atomic formula.

    c) If φ is a formula then $\neg\varphi$ is an atomic formula.

    d) If φ is a formula of the form $p \wedge q$ and connective $\wedge$ occurs in the scope of a number of existential quantifiers of the form $(\exists y_1 .... \exists y_n)$, then $(\exists y_1 .... \exists y_n)\, p(\wedge/\exists y_1 .... \exists y_n)q$ is a formula.

    e) If φ is a formula of the form $p \vee q$ and connective $\vee$ occurs in the scope of a number of universal quantifiers of the form $(\forall y_1 .... \forall y_n)$, then $(\forall y_1 .... \forall y_n)\, p(\vee/\forall y_1 .... \forall y_n)q$ is a formula.

    f) If φ is a formula and $(\exists x)$ occurs in the scope of a number of universal quantifiers of the form $(\forall y_1 .... \forall y_n)$, then $(\forall y_1 .... \forall y_n)\,(\exists x/\forall y_1 .... \forall y_n)\varphi$ is a formula.

---

[8] As mentioned in the introduction, further development of Pierce's idea was made in the late Fifties by Henkin.

g) If φ is a formula and $(\forall x)$ occurs in the scope of a number of existential quantifiers of the form $(\exists y_1.... \exists y_n)$, then $(\exists y_1.... \exists y_n)$ $(\forall x/\exists y_1.... \exists y_n)\varphi$ is a formula.

h) Nothing else is an IF formula.

One remark is in order. Hintikka did not introduce clauses d) and g)—in his presentation, only an existential quantifier or a disjunction can be independent from previous universal quantifiers in whose scopes it lies. Here the independence of the universal quantifier and the conjunction is allowed as well,   following Dechesne (2005), in order to render the syntax symmetrical. This is not an essential change because every IF formula with a slashed universal quantifier or conjunction can be transformed into an IF formula without it, as I will show later. However, there are some reasons to prefer one syntax to the other, which will become clear later on.

Now, in clauses d), e), f), and g) the set of quantifiers behind the slash sign can be empty. In that case, the usual first-order quantifiers $\exists x$, $\forall x$ or connectives $\wedge$, $\vee$ are at stake. Classical first-order expressions are special cases or the shorthand of IF expressions.

There are some changes to be made in *Definition 5* as well for the free occurrence of variables for quantifiers:

Definition 5*:

> a) Free $(\infty\varphi) = $ Free (φ);
>
> b) Free $(\varphi \wedge \psi) = $ Free (φ) U Free (ψ); Free $(\varphi \vee \psi) = $ Free (φ) U Free (ψ);
>
> c) Free $(\exists x/\forall y_1....\forall y_n)\varphi = ($ Free $(\varphi)-\{x\})$ U $\{y_1....y_n\}$;
>
> d) Free $(\forall x/\exists y_1....\exists y_n)$ $\varphi = ($ Free $(\varphi)-\{x\})$ U $\{x_1....x_n\}$.

I will add one more definition, that of a *subformula*:

Definition 10:

> The set of subformulas Sub (φ) of an IF formula φ is the smallest set S such that:
>
> a) $\varphi \in S$;

b) If $\neg\psi \in S$ then $\psi \in S$;

c) If $\psi_1(\vee/ \forall y_1.... \forall y_n) \psi_2 \in S$ then $\psi_1, \psi_2 \in S$; If $\psi_1(\wedge/\exists y_1.... \exists y_n) \psi_2 \in S$ then $\psi_1, \psi_2 \in S$;

d) If $(\exists x/\forall y_1....\forall y_n) \psi \in S$ then $\psi \in S$; If $(\forall x/\exists y_1....\exists y_n) \psi \in S$ then $\psi \in S$.

If there is more than one occurrence of the same element in a formula φ, (for example $\varphi = \psi \vee \neg\psi$), every occurrence must be treated as a different subformula. The reason for that will be clear when semantic games are introduced. Every such subformula produces a different subgame.

In Hintikka's (1996a) presentation of IF logic, it is assumed that all formulas are in the *negation normal form*, meaning that all negation signs occur in front of atomic formulas. This is a restriction introduced for the sake of simplicity, and is not a large restriction, because for every IF formula there is an equivalent IF formula in negation normal form.[9] However, following Dechesne (2005), I think that it is better to leave out this assumption, even if it renders things technically a little more complicated, because of the treatment of negation in GTS. I will make a remark about it in the following chapter.

I want to introduce one more definition of a *regular* IF formula:[10]

Definition 11: *Regular* IF formulas are IF formulas meeting two conditions: 1) if a quantifier is slashed it has to be subordinate to the quantifier of which it is independent; 2) there is no double quantification—two nested quantifiers cannot carry the same variable. Therefore, formulas $\forall x (\exists y/\forall x) \varphi$ and $(\forall y_1....\forall y_n) \varphi (\vee/ \forall y_1....\forall y_n) \psi$ are examples of regular formulas, while $(\exists y/\forall x) \varphi$ and $\forall x (\varphi \wedge \exists x \psi)$ are not.

I have now all the necessary elements of syntax, so I can move on to *Game Theoretic Semantics*, the semantics that Hintikka proposed for IF logic as its most natural semantics. Some alternative semantics have also been proposed for IF

---

[9] De Morgan's laws are holding for IF logic, as I will prove later, so they are used to push the negation signs inwards.

[10] See Caicedo, Dechesne, and Janssen (2009). This definition is important for preventing some undesirable effects in IF logic such as the *signalling*. I will discuss it in section *I.6.1*.

logic, such as a version of compositional semantics constructed by Hodges[11] and Skolem semantics. These have all been proven to be equivalent.[12]

I.3. Game theoretical semantics (GTS)

A recursive definition of truth à la Tarski relies on the concept of valuation. It satisfies compositionality, moving from the inside-out (from atomic formulas to complex ones). The truth is relative to a model and a valuation function, which attributes a certain type of entity to every constant and variable in a given language. This definition of truth, however, must be expressed in a higher-order language, that is, in a more powerful language where we can quantify over valuations. As per Tarski's famous result from 1936, it is not possible to define the concept of truth for any formal arithmetic using the expressive means of that arithmetic alone. Despite Tarski's result, Hintikka thinks that attempts to seek a realist definition of truth for the language of mathematics in the language itself might not be in vain, but this cannot be done while respecting compositionality. The semantic theory that attributes the meaning "from the inside-out" is not adequate for IF logic. Rather, the reverse method is much more promising—where meaning is obtained through a semantic game between two players, *Verifier* and *Falsifier*,[13] starting from the entire sentence and descending to its atomic expressions, whose authenticity is then checked in the model. According to Hintikka, compositionality and the "inside-out" approach amount to the fact that the sentence is considered independently of the context in which it is stated. The meaning of the sentence depends only on the meaning of its constituent parts and its structure. Different semantics, in which we start with the entire sentence until we reach its basic components, allows taking into account dependence on context.[14]

Let me repeat that the game theoretic approach was born in 1960 with *Dialogical logic*, developed by Paul Lorenzen and Kuno Lorenz.[15] As I mentioned in the introduction, this approach led to an epistemic turn that was initiated by the developement of proof-theory which in dialogues was combined with the notion of

---

[11] Although, it is not Tarskian–type semantics, see section *I.5.13.*
[12] See Mann, Sandu, Sevenster (2011).

[13] In his earlier papers, Hintikka named the players *Myself* and *Nature.*
[14] See Hintikka, Sandu (1997).
[15] The main original papers are collected in (Lorenzen, Lorenz,1978). Other papers have been collected more recently in (Lorenz, 2010a, b).

a game. In this way the dynamic features of reasoning were obtained. The basic idea of the dialogical approach to logic is that the meaning of the logical constants is given by the formal rules for their use, clearly inspired by Witgenstein's motto of *meaning as use*. This approach is an alternative to both model-theoretic and proof-theoretic semantics.[16]

In 1968 Jaakko Hintikka combined model-theory with game-theory to develop what is now known as Game Theoretical Semantics (GTS). The common feature with the dialogical framework is that the concept of truth or that of validity is based on game-theoretic concepts, such as the existence of a winning strategy for a player. However, differently to the dialogical framework, in GTS satisfaction and the validity are related to the notion of model.[17] Hintikka claims that his semantic games are exact codifications of language games in Wittgenstein's sense.

The central idea is that a language gains meaning only within certain actions of participants in a language practice. Particularly interesting is the meaning of quantifiers.[18] Activities in which their meaning is obtained can be understood as a search for (and a discovery of) the objects (or "individuals-witnesses") that the quantifiers range over. This game of "seeking and finding" is formalized as a semantic game for quantifiers. In general it is a game between the *Verifier*, who proposes the sentence and tries to defend it, and the *Falsifier*, who challenges it. Hintikka considers GTS to be the most natural semantics for IF logic.

The game is defined as follows:

Definition 12:

> Let Eloise and Abelard be the players in a game. Eloise is the initial Verifier, trying to defend the sentence at stake, and Abelard is the initial Falsifier, trying to deny it. A semantic game G (φ) for the sentence φ begins with φ. The game is played in respect to the model M with a given language L. Through various stages of game, players will consider either the sentence φ or the sentence φ_1 obtained from φ through the game. The game is played following some well-defined rules:
>
> R $\lor$-disjunction rule: game G ($\varphi_1 \lor \varphi_2$) starts with the choice of a player in the role of Verifier for φ_i (i = 1 or 2). The game continues as G (φ_i).

---

[16] See also Ruckert (2011).
[17] See Hintikka (1962, 1973, 1996a), Hintikka, Sandu (1997). See also Hintikka (1999) and in particular Hintikka et al. (1999). For the relation between the dialogical logic and GTS, see Rahman, Tulenheimo (2006).
[18] See Hintikka (1968).

R ∧-conjunction rule: game G ($\varphi_1 \wedge \varphi_2$) starts with the choice of a player in the role of Falsifier for φ_i (i = 1 or 2). The game continues as G(φ_i).

R ∃-existential quantifier rule: game G (∃x Sx) starts with the choice of a player in the role of Verifier of one member from the domain for x. If the individual is *a*, the game is played as G (S*a*).

R∀-universal quantifier rule: game G (∀xSx) starts with the choice of a player in the role of Falsifier of one member from the domain for x. If the individual is *a*, the game is played as a G (S*a*).

R~-negation rule: game G (~φ) is played the same as game G (φ), except that the players change their roles.

R-atomic sentences rule: if A is an atomic sentence that is true, the Verifier wins. If the sentence is false, the Falsifier wins.

Each application of the rules eliminates one logical constant, so in a finite number of steps we come to the rule for atomic sentences. The truth of an atomic sentence is determined in the model M, in which G (φ) is played. This is allowed by the interpretation of all non-logical constant terms in the model. The interpretation is an integral part of the model and it provides the meaning of primitive symbols in a given interpreted first-order language.

In Hintikka's (1996a) presentation, he speaks directly of the Verifier and Falsifier. Once again, following Dechesne (2005), I wanted to give a more general presentation, so I introduced Eloise and Abelard as players in a game and I assigned them the roles of Verifier and Falsifier. As Dechesne (2005:36) points out, this presentation brings to the fore the importance of the negation rule as a switch of roles of players in a semantic game. That is, Hintikka initially takes sentences to be in a negation normal form, so the change of roles of the players in the game does not really happen. As the negation sign appears only in front of atomic expressions, for negated expressions there is no play—their status is determined directly by the model.[19] In a more general presentation, the rule of negation is shown in all its importance.

Now, we can state the truth and falsity conditions for an IF sentence.

---

[19] Also recall the *definition 3* c*: if φ is an atomic formula the ¬φ is also considered to be an atomic formula.

Definition 13:

    a) Sentence φ is true in model $\mathbb{M}$ ($\mathbb{M} \models_t \varphi$) if and only if there is a winning strategy for Eloise in the game G (φ) played in $\mathbb{M}$.

    b) Sentence φ is false in model $\mathbb{M}$ ($\mathbb{M} \models_f \varphi$) if and only if there is a winning strategy for Abelard in the game G (φ) played in $\mathbb{M}$.

One more thing should be noted. The players in a game are idealised agents who are always capable of finding a winning strategy if there is one, so the game does not depend on the intellectual capacity of players. What counts is the mere existence of certain functions in the model, yielding victory for a player.

I.4. Game-theory framework

I have presented game theoretic semantics according to Hintikka.[20] A more precise formalisation is given in Mann, Sandu, and Sevenster (2011) and Dechesne (2005). I will rely on those two works in the following section. The semantic games at stake are classified in the literature as *extensive, non-cooperative, zero-sum games with imperfect information.*

I.4.1. Strategic and extensive games

Semantic games are classified as *extensive* as opposed to *strategic* games.[21] In a strategic game players decide on their movements at the beginning of the game and then the outcomes are revealed. To define the game one needs to specify a list of players, a set of possible actions every player can adopt, and players' preferences over possible outcomes. The players choose their strategies simultaneously, without knowing which strategies are chosen by their opponents;

---

[20] I have introduced a few changes compared to Hintikka's presentation with remarks about them.

[21] See Mann, Sandu, Sevenster (2011: 9).

as such, these are essentially games with *imperfect information*. A famous example is the Prisoner's Dilemma.[22]

Extensive games, on the other hand, offer a dynamic approach. In these games players move one after another so they choose their actions at every step of game. The example of an extensive game would be the game of chess.

The game is *non-cooperative* or *strictly competitive* in the sense that players don't have an interest in cooperate with one another. It is a zero-sum game because the sum of payoffs for players is constant and remains at zero, so if one player wins the other must lose.

This brings us to the important notion of *information*, which is central for IF logic. Semantic games for classical first-order logic are extensive games with perfect information. Semantic games for IF logic are extensive games with imperfect information. Let me start with games with perfect information.

I.4.2. Extensive games with perfect information

Definition 14:

An extensive game with perfect information is defined by specification of the following elements:

a) P, the set of players.

b) H, the set of *plays* or *histories* that are finite sequences. Let $(h_1...h_i)$, the elements of H, be the *initial sequence*. Then for the sequence $(h_1...h_n) \in H$, and for the sequence $(h_1...h_m) \in H$, such that $i < m < n$, we say that $(h_1...h_n)$ is an *extension* of $(h_1...h_m)$. *The terminal history* is the

---

[22] A prisoner dilemma is a story of two criminals arrested and investigated separately in two different cells. The police have enough evidence to convict them of a minor offence but not enough to convict them of a serious offence, unless some of them talks. Each of them has two options: to cooperate with a police or to plead innocent. If they both plead innocent and deny the crime, they both get one year in prison for a minor offence. If one cooperates with the police and the other does not, the one who cooperates goes free as a witness against the other and the one who does not cooperate gets four years in prison. If they both cooperate they both get three years. The optimal choice in this game turns out to be that they both cooperate with the police.

sequence that has no extension in H. The set of terminal histories is denoted T. The set of non-terminal histories is denoted N.

c)*The player function* f_p, that assigns a player from the set P to every non-terminal history in N, f_p: N→P. This function shows which player is making a move.

d)The set of *actions* A = {$a_1...a_n$}. The *action* is a step from the history h = ($h_1...h_n$) to the successor history h*h = ($h_1...h_n$, h).

f)*The utility function*, f_u, that assigns a payoff to every player p in the game, f_u: T→ R^p, where R is the set of possible payoffs for the player p in the game.

Formally, the game is strictly competitive if for all h, h' from T the following holds: $f_{u1}(h) \geq f_{u1}(h')$ iff $f_{u2}(h') \geq f_{u2}(h)$.

The game is zero-sum if the sum of players payoffs is zero so there is c∈R such that $f_{u1}(h) + f_{u2}(h) = c$ and $c = 0$.

I.4.3. Strategies

Let us now have a look at the definition of *strategy*. A strategy for a player is a description of that player's move in any possible state of a game. It is a *choice function* that suggests to the player what move she should choose whenever it is that player's turn to play.

Definition 15:

For the player p let the set $H_p = f_p^{-1}(p)$ be the set of histories where it is p's turn to play. A strategy σ is defined as: σ ∈∏ (h∈H_p) A (h).

I will yet use the following notation:

Definition 16:

The set H_σ denotes the set of histories where the particular strategy σ is followed.

The set $T_\sigma = H_\sigma \cap T$ denotes the set of terminal histories where the particular strategy σ is followed.

The set $T_p = f_u^{-1}(p)$ denotes the set of terminal histories won by the player p.

The strategy σ for the player I is the *wining strategy* if, following σ, the player I wins in every terminal history, regardless of the moves made by player II.

The game G is *determined* if one of the players has the wining strategy in G. From the *Gale-Stewart theorem* it follows that every two-player, win–lose game with perfect information and finite depth is determined.[23]

Semantic games for classical first-order formulas are extensive, two-player, win–lose games with perfect information. They are determined in the sense that one of the players always has a winning strategy if the game is of finite length, which of course must be the case because the law of the excluded middle holds for classical first-order logic. Semantically, for classical first-order formulas the game for two players is transparent. Both players have perfect information about the course of the game and they can distinguish between every action they can undertake for a given history. The slash sign in IF logic brings a limitation in information available to a player in the game, so the games for IF logic are *extensive games with imperfect information*. I will turn to these now.

I.4.4. Extensive games with imperfect information

Definition 17:

An extensive game with imperfect information is defined in the same manner as an extensive game with perfect information (*definition 14*) with one additional element:

e) The equivalence relation $\approx_p$ on the set of histories H. When there is such a relation between two histories h and h', the actions for those two histories are indistinguishable for the player p. It is denoted h $\approx_p$ h'. In that case, the player cannot distinguish between two actions available to him, so A(h) = A(h').

---

[23] Gale, Stewart (1953).

Player I actually makes a move without knowing which move player II has made, so player I is playing regardless of the move of her opponent.

The concept of strategy in extensive games with imperfect information is the same as the one previously defined, except when there is a limitation on information available for player p: in that case, the two histories are indistinguishable to him and so are the strategies concerning those two histories, $\sigma(h)=\sigma(h')$.

I.4.5. Semantic games for IF logic

Let me specify a semantic game for IF logic in a given framework.

Definition 18:

If φ is an IF formula, $\mathbb{M}$ a suitable model, and v a valuation on the domain that contains Free(φ), the game is defined with a specification of the following elements:

a) P = { $\forall, \exists$ }, the set of players, consisting of two players, Abelard and Eloise.

b)H, the set of *plays* or *histories*, which are finite sequences that show possible moves in a game. The game can have only one *initial history*. Starting from the initial history, the players can either be in a position where the game-rules suggest to one of them that they should make a move (*the position of decision*), or they can be in a position where no move is available to any player, so that the game comes to its end and the payoffs are revealed (*the terminal history*, the sequence that has no extension in H). T denotes the set of terminal histories whereas N denotes the set of non-terminal histories.

In a position of decision there are two kinds of moves available to a player. If he arrives at a quantifier he chooses an element of a domain. If he arrives at a connective he chooses between subformulas in question. The order and the kind of move at stake are determined by the syntax, starting with the entire sentence and reaching its component parts. I will define the set of histories H recursively, by defining the set of finite histories H_ψ for every $\psi \in$ Sub(φ).

$H_\varphi = \{v, \varphi\}$;

If ψ is atomic the induction stops;

If $\psi = \psi_1 \vee \psi_2$ then $H_{\psi i} = \{h * \psi_i : h \in H \ \psi_1 \vee \psi_2\}$;

If $\psi = \psi_1 \wedge \psi_2$ then $H_{\psi i} = \{h * \psi_i : h \in H \ \psi_1 \wedge \psi_2\}$;

If $\psi = (\exists x / \forall y_1 .... \forall y_n) \psi_1$ then $H_{\psi 1} = \{h*(x,a) \ : \ a \in M$, $h \in H(\exists x / \forall y_1 .... \forall y_n)\}$;

If $\psi = (\forall x / \exists y_1 .... \exists y_n) \psi_1$ then $H_{\psi 1} = \{h*(x,a) \ : \ a \in M$, $h \in H(\exists x / \forall y_1 .... \forall y_n)\}$;

then $H = U \ \psi \in Sub(\varphi) H_\psi$.

Every $h \in H$ defines a valuation v_h for the expression ψ_h.

$v_{h'} = v$ for $h' = (v, \varphi)$;

$v_{h'} = v_h$ for $h' = h * \psi$;

$v_{h'} = v_h(x/a)$ for $h' = h*(x,a)$.

c)*The player function* f_p assigns a player from the set P to every non-terminal history in N, f_p: N→P. The function shows which player is making a move. If a history at a position of decision is a subformula with disjunction or existential quantifier then the player function demands that Eloise make a move. If a subformula is conjunction or universal quantifier then it is Abelard who plays.

f_p (h) = $\exists$ if h = $\psi_1 \vee \psi_2$ or h = $(\exists x / \forall y_1 .... \forall y_n) \psi_1$;

f_p (h) = $\forall$ if h = $\psi_1 \wedge \psi_2$ or h = $(\forall x / \exists y_1 .... \exists y_n) \psi_1$.

d) *The set of actions* A = $\{a_1 ... a_n\}$. The action *a* is a step from the history h = $(h_1 ... h_n)$ to the successor history h*h = $(h_1 ... h_n, h)$.

f) *The utility function*, f_u which assigns a payoff to a player *p* in the game, f_u: T→ R^p, where R is a set of possible payoffs for the player *p* in the game. When atomic formulas are reached the game ends and the outcomes are revealed. If the atomic expression ψ_h reached at the end is satisfied in the model under given valuation then Eloise wins, otherwise the winner is Abelard.

$f_u(h) = \exists$ if ψ_h is satisfied in $\mathbb{M}$ under v_h;

$f_u(h) = \forall$ if ψ_h is not satisfied in $\mathbb{M}$ under v_h.

e) *The equivalence relation* $\approx_p$ on the set of histories H, which provides the meaning of the slash sign. Such a relation means that actions for those two histories are indistinguishable for the player p, or h $\approx_p$ h', so A(h) = A(h').

Informally, when a player is to make a move for a subformula with the slash sign he must do it without knowing the values of the variables under the slash sign that have already been chosen.

If ψ is not atomic and $\psi \in Sub(\varphi)$ then let Y denote a set of variables occurring under the slash sign in ψ (the set can be empty). Let V be the set of valuations for ψ. Then two valuations v and v' in V are equivalent if they assign the same values to variables in ψ except for variables in Y, or

$v \approx v'$ iff $v(x) = v'(x)$ for all $x \in Free(\varphi)- Y$

The equivalence relation between two valuations produces the equivalence relation on the set of histories in ψ, H_ψ.

For h,h' $\in H_\psi$: h $\approx_p$ h' iff $v_h \approx v_{h'}$.

It should be noted that the restriction in information available to a player in a semantic game does not forbid the player from performing any action. It rather concerns possible strategies available to him. Therefore, there is no specific game rule for the slash sign. The slash is not another propositional operator, rather it is a restriction in information embedded at the object-language level in order to make a selection between winning strategies.

Definition 19:

For the player p let the set $H_p = f_p^{-1}(p)$ be the set of histories where it is p's turn to play. The strategy σ is defined as $\sigma \in \prod (h \in H_p)$ A (h).

Definition 20:

The set H_σ denotes the set of histories where the particular strategy σ is followed.

The set $T_\sigma = H_\sigma \cap T$ denotes the set of terminal histories where the particular strategy σ is followed.

The set $T_p = f_u^{-1}(p)$ denotes the set of terminal histories won by p.

I have already defined the satisfaction of IF formulas in a model. To complete the presentation I will state it again.

Definition 13:

a) An IF formula φ is true in the model $\mathbb{M}$ ($\mathbb{M} \models_t \varphi$) if and only if there is a winning strategy for Eloise in the game $G(\varphi)$ played in the model $\mathbb{M}$.

b) An IF formula φ is false in the model $\mathbb{M}$ ($\mathbb{M} \models_f \varphi$) if and only if there is a winning strategy for Abelard in the game $G(\varphi)$ played in the model $\mathbb{M}$.

Let me present now some examples of games with perfect and with imperfect information starting with an expression without a slash.

Example 1:

Let φ be $\forall x \exists y \, (x = y)$ and the model $\mathbb{M} = \{0,1\}$.

The initial history is $h = (\emptyset, \varphi)$. As the formula starts with the universal quantifier the player function asks Abelard to make the first move and to choose the value for x. He has two possibilities, $h_1 = (\emptyset, \varphi, (x,0))$ and $h_1' = (\emptyset, \varphi, (x,1))$. The next step brings in the existential quantifier, so it is Eloise who plays and chooses the value for y. As there is no slash sign, Eloise has information about the value Abelard has chosen for x and all the histories are distinguishable for her. Her strategy is simple—to choose the same value for y as Abelard has chosen for x.

$\sigma(h_1) = (y, 0)$; $\sigma(h_1') = (y, 1)$.

Therefore, we have two terminal histories: $h_2 = (\emptyset, \varphi, (x,0), (y, 0))$ and

$h_2' = (\emptyset, \varphi, (x,1), (y, 1))$. In both of these Eloise has a wining strategy, so we conclude that φ is satisfied in $\mathbb{M}$.

Example 2:

Let φ be $\forall x(\exists y/\forall x)\,(x = y)$ and the model $\mathbb{M} = \{0,1\}$.

The initial history is the same as before: $h = (\emptyset, \varphi)$. We start with the universal quantifier, so first it is Abelard's turn to play. Again, he has two possibilities: $h_1 = (\emptyset, \varphi, (x,0))$ and $h_1' = (\emptyset, \varphi, (x,1))$. Then it is Eloise who is in the position of decision but this time the existential quantifier is independent of the universal quantifier and she has to make a choice for y without knowing the value Abelard has chosen for x. Two histories h_1 and h_1' are indistinguishable for her: $h_1 \approx_\exists h_1'$. Her strategy σ cannot take as argument the value of x, so it must be a constant function. $\sigma\colon (h_1) = \sigma\,(h_1') = (y, m);\ m = \{0,1\}$

Terminal histories are either:

$h_2 = (\emptyset, \varphi, (x,0), (y, 0))$ and $h_2' = (\emptyset, \varphi, (x,1), (y, 0))$ or

$h_2{}^* = (\emptyset, \varphi, (x,0), (y,1))$ and $h_2'{}^* = (\emptyset, \varphi, (x,1), (y, 1))$.

In either case, Eloise does not win both terminal histories and consequently she does not have a winning strategy. Therefore, φ is not satisfied in the model. But it is clear immediately that Abelard does not have a wining strategy either—he cannot win both terminal histories. The law of the excluded middle does not hold for IF logic. I will investigate this separately. The Gale-Stewart theorem does not hold for extensive games with imperfect information.

I.5. Properties of IF logic

I will proceed by discussing some important properties of IF logic.

I.5.1. Notion of strategy

As presented priviously, GTS truth conditions for a sentence are defined in terms of strategies for two players who are trying to verify or falsify the sentence. The truth is defined, in Wittgenstein's manner, through the practice of verification and falsification within the given rules.

The strategy is expressed by the finite set of choice-functions or *Skolem functions* whose values suggest what individuals the Verifier has to choose in his actions related to the existential quantifier and disjunction in order to win the game. The same holds for the Falsifier's strategy, which is related to the universal quantifier and conjunction. Those functions belong to the existential part of second-order language, ususally denoted Σ^1_1.[24] I will show later that the expressive power of IF logic is exactly that of the Σ^1_1 second-order fragment. The truth conditions for a sentence φ can be expressed in the Σ^1_1 fragment. Such a translation is always feasible. Here is an example:

1)$\forall x \exists y\ S(x,y)$

The sentence 1) is true if there is a winning strategy for the Verifier. The Verifier's strategy shows her how to select y in function of the value of x, $(f(x) = y)$ in order to win the game. The expression of the existence of such a strategy is the following:

2) $\exists f \forall x\ S(x,f(x))$

I.5.2. The axiom of choice

If 1) and 2) are linked with a conditional, the following formulation of the *axiom of choice* is obtained:

3) $\forall x \exists y\ C(x,y) \rightarrow \exists f \forall x\ C(x,f(x))$

From the GTS point of view, the truth of 3)—which expresses one version of the axiom of choice—is in fact derived from the very definition of truth. In fact, as I will discuss later, it is related to the truth of the universal.

Let me point out that the axiom of choice is essential for GTS, since without it the Tarski-style semantic and GTS for classical first-order logic would not be equivalent. The reason for this is that strategies in GTS are understood as

---

[24] The existential second-order logic is the fragment of second-order logic that consists of formulas in the form $\exists x_1... \exists x_n \Psi$, where $\exists x_1... \exists x_n$ are second-order quantifiers and Ψ is a first-order formula.

deterministic strategies that impose choices on the Verifier and the Falsifier, leaving no real options. Wilfried Hodges (2013) sees this as a weakness in Hintikka's approach and states that a more natural way to conceptualise GTS would be to use *non-deterministic* strategies so that no a priori presumption of the axiom of choice is needed. However, Hintikka insists that there is nothing troubling about the axiom of choice and that it actually constitutes our conception of truth.

> *This paradigm problem concerns the status of the axiom of choice. This axiom was firmly rejected by Brouwer and it was mooted in the controversies between the French intuitionists and their opponents....The axiom of choice is true. The idea of "choosing" or "finding" suitable individuals is systematised in what is known as game-theoretical semantics. For mathematicians, this semantics is no novelty, however, but little more than a regimentation and generalisation of the way of thinking that underlies mathematicians' classical (or perhaps I should say Weierstrassian) epsilon-delta analyses of the basic concepts of calculus, such as continuity and differentiation...(2001:8).*

> *...To return to the usual axiom of choice, it is thus seen to be unproblematically true. How can any intuitionist deny the axiom of choice...? What can possibly go wrong here? Moreover, evoking the concept of knowledge, either in the form of epistemic logic or informally, does not seem to help an intuitionistic critic of the axiom of choice at all, either....(2001:9)*

> *...The discussion of the axiom of choice between intuitionists and classicists has conducted at cross-purposes. It can only be dissolved by making distinction between knowing that and knowing what that neither party has made explicit (2001:13).*

> *Zermelo did not begin to axiomatize set theory unselfishly from the goodness of his theoretical heart. His main purpose was to justify his well-ordering theorem. In practice, this largely meant to justify the axiom of choice. [...]. But that is not the full story. Worse still: Zermelo's specific enterprise was unnecessary, in that the so- called axiom of choice turns out to be in the bottom a plain first-order logical principle.* (Hintikka, 2011: 13)

Hintikka argues that GTS justifies the axiom of choice in a way perfectly acceptable for constructivists. Moreover, as per the last sentence in the previous quotation, the axiom of choice should be considered a first-order principle. In fact, as I intend to show in Chapter III, the latter is true: the axiom of choice involves only first-order logic. As for the former, I will argue below that the axiom is indeed acceptable for constructivists only if the intensionality of the choice function is assumed. However, Hintikka wants a logic that can render classical mathematics, but this is only possible if extensionality is presumed. Having both the extensionality and the axiom of choice—without assuming the unicity of function—does not seem to be possible.

I will address this issue and discuss it in more detail later. I will also compare Martin-Löf's analysis of the axiom of choice with Jaakko Hintikka's stance on it. As I mentioned in the introduction, my aim is to show that dependence and independence features that are essential in IF logic can be formulated within the framework of Constructive Type Theory (CTT), but without the shortcomings: without sacrificing the full axiomatization of a system and without giving up on the theory of inference.

The framework of dialogical logic offers a very subtle view on the issue. Recent developments in dialogical logic show that the CTT approach to meaning in general and to the axiom of choice in particular is very natural to game-theoretical approaches where metalogical features are explicitly laid out in object-language. I will argue for the fruitfulness of a game-theoretical semantics in the context of the foundation of mathematics, thus confirming Hintikka's basic idea. However, in my opinion, the game-theoretical framework must be constructed in a rather different way.

I.5.3. Skolemisation procedure

I will present in this section the skolemisation procedure for IF logic. I start off with a classical first-order formula φ in the negation normal form. The Skolem form of φ is obtained by the following transformations:

If $\exists y(\psi) \in Sub(\varphi)$ and it lies within the scope of a string of universal quantifiers $\forall x_1....\forall x_n$, take the n-ary function symbol f, erase the existential quantifier $\exists y$ and replace all occurrences of y in ψ with $f(x_1...x_n)$. Do the same with all the existential quantifiers in φ until obtaining a first-order formula φ', in which all the functional symbols $f_1...f_n$ occur with the right arguments and there is no existential quantifier. In front of φ' add the string $\exists f_1...\exists f_n$. In this manner, a Σ^1_1 formula is obtained in the form $\exists f_1...\exists f_n(\varphi')$. Below is an example of the first-order formula presented above:

Example 3 :

$$\forall x \exists y \, S(x,y) \rightarrow \exists f \forall x \, S(x,f(x))$$

The Skolem form of an IF sentence is obtained in the same manner, with some changes when a slashed existential quantifier occurs in the form $(\exists y/\forall x_1....\forall x_n)$. Then the existential quantifier should be deleted, but this time the variables $(x_1...x_n)$ are not the arguments of a newly introduced function, because the existential quantifier is independent of the universal quantifiers under the slash sign.

Example 4:

$$\forall x_1 \forall x_2 \, (\exists y/\forall x_1) \, S(x, y) \rightarrow \exists f \forall x_1 \forall x_2 \, S(x, f(x_2))$$

In addition, we need to provide the Skolem form of disjunction, because disjunction is also a decision position for Eloise. Let φ be a classical first-order formula in the negation normal form, where disjunction $\psi_1 \vee \psi_2 \in Sub(\varphi)$ occurs in the range of a string of universal quantifiers of the form $\forall x_1....\forall x_n$. Then choose a new n-ary functional symbol f with $(x_1...x_n)$ as arguments and $\{0,1\}$ as its range. That implies that there are at least two elements in a model that are interpretations of the constants 0 and 1. Replace $\psi_1 \vee \psi_2$ with

$$(f(x_1...x_n) = 0 \wedge \psi_1) \vee (f(x_1...x_n) = 1 \wedge \psi_2).$$

The first disjunct corresponds to Eloise's choice of the left disjunct and the second to her choice of the right disjunct in the original formula. Do the same with every disjunction in the formula, thereby obtaining a new formula φ'. As previously explained, in front of φ' one should add the string $\exists f_1...\exists f_n$ to obtain a Σ^1_1 formula in the form $\exists f_1...\exists f_n(\varphi')$.

Example 5:

$$\forall x_1 \forall x_2 (P(x_1, x_2) \vee R(x_1, x_2)) \rightarrow \exists f \forall x_1 \forall x_2 \, (f(x_1, x_2) = 0 \wedge P(x_1, x_2)) \vee$$

$$(f(x_1, x2) = 1 \wedge R(x_1, x_2)).$$

If φ is an IF formula where a disjunction is independent of a string of universal quantifiers in the form $\forall x_1....\forall x_n$, the procedure is the same, only the function symbols do not take as arguments variables $x_1...x_n$, occurring under the slash sign. Here is an example.

Example 6:

$$\forall x_1 \forall x_2 \, (P(x_1, x_2) \lor / \forall x_1 \, R(x_1, x_2)) \rightarrow \exists f \forall x_1 \forall x_2 \, (f(x_2) = 0 \land P(x_1, x_2)) \lor (f(x_2) = 1 \land R(x_1, x_2)).$$

Let me make a few remarks. We can see now how the skolemisation procedure yields the truth conditions for an IF sentence and why the axiom of choice is important in GTS. By GTS rules, existential quantifiers and disjunctions are decision positions for Eloise (i.e. the initial Verifier) in a game for a given sentence. If Eloise has a wining strategy in the game the sentence is true. When the IF sentence is transformed to its Skolem form, the existential quantifier and disjunction disappear and instead we explicitly see the choice functions that are guiding every one of Eloise's moves. The arguments of those functions show the moves of Abelard that Eloise's choices depended on. Here is the formal definition of Skolem form of an IF sentence:

Definition 14:

If φ is an IF sentence of vocabulary L, and $\mathbb{M}$ is a model with at least two elements that provide interpretations of constant symbols 0 and 1 that are not in L, the Skolem form of φ, Sk (φ) is obtained with the following rules:

sk (φ) [$R(t_1, \ldots, t_n)$] = $R(t_1, \ldots, t_n)$.

sk (φ) [$\sim R(t_1, \ldots, t_n)$] = $\sim R(t_1, \ldots, t_n)$.

sk (φ) [$\psi_1 \, (\lor / \, \forall x_1 \ldots \forall x_n) \, \psi_2$] = (Sk ($\varphi$) [$\psi_1$] $\land f(y_1 \ldots y_n) = 0$) $\lor$ (Sk (φ) [ψ_2] $\land$

$f(y_1 \ldots y_n) = 1$); where f is a new function symbol and the variables $y_1 \ldots y_n$ are bound by a string of universal quantifiers $\forall y_1 \ldots \forall x_n$ preceding $\psi_1 \, (\lor / \, \forall x_1 \ldots \forall x_n) \, \psi_2$ in φ, but not among $x_1 \ldots x_n$.

sk (φ) [$\psi_1 \, (\land / \exists x_1 \ldots \exists x_n) \, \psi_2$] = sk ($\varphi$) [$\psi_1$] $\land$ sk (φ) [ψ_2].

sk (φ) [$(\exists y / \forall x_1 \ldots \forall x_n)\psi$] = sk ($\varphi$) [$\psi$] [$y$ replaced with $f(x_1' \ldots x_n')$] where f is a new function symbol and the variables $x_1' \ldots x_n'$ are bound by a string of universal quantifiers $\forall x_1' \ldots \forall x_n'$ preceding $(\exists y / \forall x_1 \ldots \forall x_n)\psi$ in φ, but not among $x_1 \ldots x_n$.

sk (φ) [$(\forall y / \exists x_1 \ldots \exists x_n) \, \psi$] = $\forall y$ sk (φ)[ψ].

As previously mentioned, in Hintikka's presentation of syntax for IF logic only existential quantifiers and disjunctions can appear slashed. Hintikka places emphasis on truth conditions that concern Eloise's moves in a game. Indeed, every IF sentence with slashed universal quantifiers and conjunctions can be transformed into the sentence without it. However, I follow the authors,[25] who think that the falsity conditions for IF sentences are equally important. As such, I have presented a larger syntax, where slashed universal quantifiers and slashed conjunctions are also allowed. This is justified by the fact that the law of the excluded middle does not hold in IF logic.

I.5.4. Kreisel counterexamples

Wining strategies for Eloise are captured by the existence of Skolem functions. In the same manner, wining strategies for Abelard are captured by the existence of Kreisel counterexamples. A sentence in the Kreisel form provides the falsity conditions for the corresponding IF sentence (see Mann, Sandu, and Sevenster, 2011: 75). The procedure of transforming an IF sentence φ into the Kreisel form is called *kreiselization* and it is actually the skolemisation of a dual sentence $\sim \varphi$.

Theorem:

$$M \models_{f} \varphi \text{ iff } M \models_{t} \sim\varphi$$

The proof follows immediately from the rule of negation as an exchange of roles for players in GTS.

The Kreisel form of IF sentences is defined in the following manner:

Definition 15:

If φ is an IF sentence of vocabulary L, and M is a model with at least two elements that serve as interpretations of constant symbols 0 and 1 that are not in L, the Kreisel form of φ (kr (φ)) is obtained with the following rules:

kr (φ) [$R(t_1,\ldots, t_n)$] = $\sim R(t_1, \ldots, t_n)$.

---

[25] See Mann, Sandu, Sevenster (2011) and Dechesne (2005).

$$kr\,(\varphi)\,[\psi_1 \lor \psi_2] = kr(\varphi)\,[\psi_1] \land kr(\varphi)\,[\psi_2]$$

$$kr\,(\varphi)\,[\psi_1 \land \psi_2] = kr(\varphi)\,[\psi_1] \lor kr(\varphi)\,[\psi_2]$$

$$kr\,(\varphi)\,[(\exists y/\forall x_1....\forall x_n)\psi] = \forall y\;kr\,(\varphi)\,[\psi]$$

$kr\,(\varphi)\,[(\forall y/\,\exists x_1...\exists x_n)\,\psi] = kr\,(\varphi)\,[\psi]$ [y replaced with $f(x_1'...x_n')$] where f is a new function symbol and variables $x_1'....x_n'$ are bound by a string of existential quantifiers $\exists x_1'...\exists x_n'$ preceding $(\forall y/\,\exists x_1...\exists x_n)\,\psi$ in φ, but not among $x_1...x_n$.

I.5.5. The law of the excluded middle

The failure of the law of the excluded middle in IF logic is due to the possibility of imperfect information in a game. I remarked before that the Gale-Stewart theorem does not hold for extensive games with imperfect information, so those games are not always determined. I have already presented the example of an IF formula that is neither true nor false ($\varphi = \forall x(\exists y/\forall x)\,(x = y)$ and the model $\mathbb{M} = \{0,1\}$). It is interesting to reflect on the nature of non-determinacy. Are we talking about the third truth value?

Hintikka does not think that there is third truth value involved. According to him, the non-determinacy indicates that not all types of choice functions exist in a given model; it is thus a structural characteristic of the model. The argument is provided in (Tulenheimo, 2009, note 27.):

> *No third truth-value is stipulated in the semantics of **IFL**. Truth-value gaps may arise when neither truth nor falsity can be meaningfully ascribed to a sentence (because a presupposition of the sentence is not satisfied). This is not what non-determinacy means. Non-determinacy is a model-relative complex negative property, which can be ascribed to a sentence. A sentence has this property iff neither of the simple positive properties of truth and falsity can be correctly ascribed to the sentence. An ascription of non-determinacy is correct or incorrect depending exclusively on the model relative to which the ascription is effected.*

However, some other authors speak about the third truth value (Pantasar, 2009, Mann, Sandu, and Sevenster, 2011).

Let me compare the failure of the law of the excluded middle in IF logic with the case of intuitionstic logic. In intuitionism the reason for the failure of the law is epistemological. It concerns the ability to construct a proof for a given formula. The case is different in IF logic. Following Hintikka, the non–existence of a certain function is a structural property of the model in question, so we are on ontological level instead. In Hintikka's own words (Hintikka, 2002a: 589):

> *Secondly, the failure of the excluded middle is purely structural, combinatorial matter. Weather or not (S ∨~S) is true depends on whether there exists a winning strategy for either player in G(S). And this depends only on the structure of the model ("possible world") on which G(S) is played, i.e. has nothing to do with the limitations of human knowledge or the knowledge of any particular human being.*

This might be a reason to speak about the third truth value in IF logic after all.

Considering the failure of the law of the excluded middle, Hodges (2013) remarks that the rule for dual negation, as a switch of roles of players in a game, is no longer justified in IF logic. According to this view, the rule for dual negation makes sense only in classical first-order logic where the law holds. Hintikka's brief direct response (2006a) was that no justification is needed because the rule for dual negation captures the very intuitive meaning of negation in game theory. The justification of dual negation can be found in (2002a): IF logic is an extension of classical first-order logic with a number of classical properties—De Morgan's laws, interdefinability of quantifiers, double negation law, distribution laws, and metatheorems such as compactness, the Löveinheim-Skolem theorem, the separation theorem, and so on. A classical first-order sentence is understood as a special case of an IF sentence—one where the set of variables under the slash sign is empty. In Hintikka's own words:

> *IF logic is thus reached not by changing anything in the classical (read: old-fashioned) first - order logic, but by liberalising it in the spirit of the original enterprise.* (Hintikka, 2002a: 589)

From this point of view, all the game rules for classical first–order logic must remain the same in IF logic. According to Hintikka (2002a), IF logic should not be considered non-classical but rather *hyperclassical* logic.

It should be noted that the dual negation in front of atomic formulas behaves in the same way as classical negation—atomic expressions are either true or false. There is a certain gap in the treatment of atomic and complex formulas on this point, due to the introduction of imperfect information in a game. Truth and falsity conditions for IF sentences can be formulated in the Σ^1_1 second-order fragment, and this fragment is not closed under classical negation. This feature is important; I will argue later that it becomes an obstacle to obtain the constructivist reading of IF logic, which Hintikka suggests. I will also argue that it is precisely under the constructivist reading that IF logic is granted the position of genuine first-order logic; but this is also where the real problem for Hintikka emerges.

Related to the law of the excluded middle, let me underline another difference between IF and intuitionistic logic. In intuitionistic logic atomic formulas might also be undetermined. In IF logic atomic sentences are always determined in the model, in the classical way. The reason for this is that in front of atomic sentences dual negation behaves as a classical, contradictory negation.

One more difference is that, unlike in the intuitionistic logic, in IF logic the law of double negation holds ($\sim\sim A \rightarrow A$). Double negation amounts to the double role-exchange of players in a row, so that the game for $\sim\sim A$ is identical to the game for A.

I.5.6. The expressive power of IF logic

In some cases, the slash sign does not contribute to anything that could not be expressed without it, but sometimes it allows for expressing the structures that would not be expressible in classical first-order logic. The classic example is the expression with Henkin's branching quantifiers. It can be represented as:

$$\left.\begin{array}{l} \forall x \exists y \\ \\ \\ \forall z \exists u \end{array}\right\} \quad S(x,y,z,u)$$

There is no linear way to express this formula in the classical first-order logic. The only means of expressing dependencies between quantifiers in classical first–order logic is the nesting of quantifiers' scopes - and those scopes cannot be arranged in such a way that they overlap only partially. The Skolem form of the previous sentence is:

$$\exists f \exists g \forall x \forall z\ S(x, f(x), z, g(z))$$

which can be translated to IF language in the following manner:

$$\forall x \forall z\ (\exists y/\forall z)\ (\exists u/\forall x)\ S(x, y, z, u).$$

In his study of branching quantifiers, Walkoe (1970) shows that the expressive power of formulas with branching quantifiers is precisely that of the existential fragment of second-order logic. Independently, Walkoe (1970) and Enderton (1970) have also shown that every existential second-order sentence Σ^1_1 is equivalent to the second-order truth or falsity condition of an IF sentence. Therefore, IF logic captures the precise expressive power of Henkin's branching quantifiers.[26]

I have already shown that every IF sentence can be translated into an Σ^1_1 sentence through the procedure of skolemisation. Now, let me describe the reverse process of the translation of Σ^1_1 sentences into IF sentences.[27]

Let Φ be an Σ^1_1 formula. Rewrite Φ in the form $\exists f_1 ... \exists f_k\ \forall x_1 ... \forall x_n\ (\varphi)$, where φ is a quantifier-free classical first-order sentence and the function symbols $f_1 ... f_k$ and variables $x_1 ... x_n$ occur in such a manner that functional symbols are not nested and that every function symbol has only one string of arguments. Let us denote the new formula Φ'. Φ' is translated into an IF sentence with the following steps: remove the string of quantified second-order function symbols $\exists f_1 ... \exists f_k$ and every $\exists f_i$ replace with a slashed first-order quantifier $(\exists y_i/\forall x_1 .... \forall x_m)$ in front of φ, where $x_1 ... x_m$ are the variables from the set $\{x_1 ... x_n\}$ that do not occur as arguments of function symbol f_i. Then in φ replace every functional symbol f_i and its arguments with y_i. In this way the IF sentence is obtained.

I have yet to show how the sentence Φ' is obtained from the sentence Φ. Let L be the language that contains the equality sign =, and $\mathbb{M}$ be the model that has at least two elements, serving as interpretations of the constant symbols 0 and 1. First we need a sentence in the form $\exists f_1 ... \exists f_k\ \forall x_1 ... \forall x_n\ (\varphi)$. So, if there is a predicate bound by an existential quantifier in Φ, we should present the predicate as its characteristic function. For example, if Φ is $\exists R \forall x\ (Rx)$, Φ' will be $\exists f \forall x\ (f(x) = 1)$. Second, if Φ is $\exists f_1\ \exists f_2\ \forall x\ (\varphi\ (f_2(f_1(x))))$, that is, if in Φ two function

---

[26] More about branching quantifiers can be found in Blass, Gurevich (1986).
[27] Hintikka (1996a: 62-63) explains the translation procedure.

symbols are nested, we introduce a new variable to put Φ' in the form: $\exists f_1 \exists f_2 \forall x_1 \forall x_2 (x_2 = f_1(x_1) \rightarrow_\varphi f_2(x_2))$. And finally, if Φ is in the form $\exists f_1 \forall x_1 \forall x_2 (\varphi(f_1(x_1), f_1(x_2))$, that is, the same function symbol has two different arguments, we transform it to Φ' by introducing a new function symbol:

$$\exists f_1 \exists f_2 \forall x_1 \forall x_2 (x_1 = x_2 \rightarrow f_1(x_1) = f_1(x_2) \wedge \varphi(f_1(x_1), f_1(x_2))).$$

Consequently, IF logic has a bigger expressive power then classical first-order logic. An example of a property of a domain $\mathbb{M}$ that can be expressed in IF logic, but not in classical first-order, is the Dedekind-infinity.

Example 7:

> The domain is Dedekind-infinite if there is an injective function from the domain to its proper subset. In the second-order logic we can express this with the following sentence:
>
> $$\exists f_1 \exists f_2 \exists m \forall x_1 \forall x_2 ((f_1(x_1) = f_2(x_2) \leftrightarrow x_1 = x_2) \wedge f_1(x_1) \neq m).$$
>
> This sentence expresses truth conditions of the IF first-order formula:
>
> $$\exists y \forall x_1 \forall x_2 (\exists y_1 / \forall x_2)(\exists y_2 / \forall x_1)(x_1 = x_2 \leftrightarrow y_1 = y_2 \wedge y \neq y_1).$$

Other properties that can be expressed in IF logic but not in classical first- order logic are: the accountability of a structure, the equicardinality of predicates of two first-order formulas, the ill-foundedness of a linear ordering, the topological notion of an open set, and so on.[28] Hence, in Hintikka's opinion this is one of the biggest advantages of IF logic: we can use it to express some important mathematical properties but without crossing over to higher-order languages. This makes IF logic the best candidate for the foundations of mathematics. Better still, IF logic has some important and desirable properties.

Beside its application in mathematics, Hintikka and his associates think that IF logic and GTS are more suitable for the purpose of a formalisation of natural language. One of the Hintikka's favourite examples of language phenomena that can be treated in a satisfactory manner with GTS (and sometimes exclusively with IF logic) is anaphora. IF logic allows taking into account different patterns of dependency among logical expressions, it is therefore more appropriate for the formalisation of natural language. Moreover, all of this can be done on the first-order level.

---

[28] see Hintikka (1996a: 186-190).

The following is an example of a natural language sentence that involves branching quantifiers, from Hintikka (1973: 344):

> *Some relative of each villager and some relative of each townsman hate each other.*

The sentence can be expressed by the existential second-order sentence:

$$\exists f\,\exists g\forall x\forall z\,((\,\text{Villager}(x) \wedge \text{Townsman}\,(z)) \rightarrow \text{Relative}\,(x, f(x)) \wedge$$

$$\text{Relative}\,(z, g(z)) \wedge \text{Hate}\,(f(x), g(z)))),$$

which has a translation in first-order IF language, but otherwise cannot be expressed at the first-order level.[29] The GTS approach together with IF logic thus provide an interpretation of such complicated examples of anaphoric expressions.

Later I will compare the GTS approach to anaphora with the interpretation found with dialogical framework for CTT. I will argue that the GTS approach has considerable advantages over other existing approaches. However, I will show that in the dialogical approach a natural interpretation of anaphoric expressions can be given without paying the price of using IF logic that is neither axiomatizable nor has an underlying theory of inference. I will give an important example that will also help me to shed some light on the way that meaning is obtained in general.

I will proceed now with the properties of IF logic.

I.5.7. The failure of axiomatisability

Kurt Gödel proved in 1930 that classical first-order logic has a complete axiomatisation, that is, it has a sound and complete proof procedure. The same property does not apply to IF logic. Its failure of axiomatisability is actually due to its great expressive power, so Hintikka does not consider this to be a big disadvantage. Hintikka (1996a: 66–67) states the following:

---

[29] This claim was objected by Gierasimczuk and Szymanik (2009). They tried to offer an alternative reading to Hintikka's examples involving branching quantifiers which makes them expressible in the first-order level with a linear representation, without using more syntactic tools then those from the classical first–order logic. However interesting their analysis might be, it is not relevant for my further discussion of branching quantifiers.

Some philosophers have seen in the absence of a complete proof procedure the basis of a serious objection to IF first-order logic. Such a way of thinking is based on a number of serious mistakes. Apparently the skeptics think that they cannot understand a language unless they understand its logic; that they do not understand its logic unless they understand what its logical truths are; and that they do not understand the logical truths of a part of logic unless they have a complete axiomatization of such truths (valid formulas). The last two steps of this line of thought are both fallacious...The subtler mistake seems to be that in order to understand logic one has to understand its logical truths. This is at best a half-truth. One possible mistake here is to think that logical truths are merely a subclass of truths. Hence, according to this line of thought, a satisfactory account of truth for a language ought to yield as a special case an account of logical truth for this language. But logical truth simply is not a species of (plain) truth, notwithstanding the views of Frege and Russell. Logical truths are not truths about this world of ours. They are truths about all possible worlds, truths on any interpretation of nonlogical constants. This notion is an animal of an entirely different color from ordinary truth in some one world. In this sense, to call valid sentences "logically true" is a misnomer. Several philosophers who swear by the ordinary notion of truth are nonetheless more than a little skeptical of the ultimate viability of the very notion of logical truth.

In the absence of a complete proof system, IF logic is inevitably linked to a model theory. Hintikka's focus is on the material truth of a model. Validity is conceived as truth in all models. I will show later that the idea behind the expressivity of IF logic is to be able to reduce the problem of deciding whether a second-order mathematical sentence is valid to the problem of the validity of its eqivalent IF sentence. However, it seems strange to propose a reduction of mathematics to a logic that lacks a proof system.[30] I will touch on this point later.

In order to justify his model-theoretic approach, Hintikka (1996a: 1–22) investigates the role of logic in mathematics. In examining this question, Hintikka points out that this role is in fact multiple and that some of its aspects are neglected. What one usually has in mind when thinking of the role of logic in mathematics is its *deductive* role or what Hintikka calls *a theory of demonstration*. Regardless of whether we deal with an interpreted or non-interpreted mathematical system, inferences of theorems from basic principles are completely logical in their nature. An axiomatized logic thus serves to secure inferences in mathematics.

---

[30] I will explain later the specific logicist programme Hintikka advocates.

Then, one can ask whether these inferences can be fully covered by computable, recursive formal rules.

Hintikka points to some other roles played by logic. One of these is often disputed, but for him the most important—which he calls the *descriptive* role—is performed by a model theory. Again, regardless of whether it is about an interpreted or non-interpreted mathematical system, logical terms are used to analyse the meaning of mathematical axioms, with the idea that the meaning is determined by a class of structures or models that can be assigned to them. Thus, we should specify the class of models M (φ) for a sentence φ, where, simply put, M is a model for φ if and only if φ is true in M. If mathematics is approached through a model theory, the problem of the definition of truth for mathematical theory comes to the fore—a question I will also address later.

The third function of logic in mathematics is its role as a medium for axiomatic set theory.

In parallel with the different roles played by logic in mathematics, Hintikka elaborates on the different types of completeness of a deductive system. Hintikka (1996a: 88–105) distinguishes several of these, among others (in his, not quite usual, terminology): *semantic, deductive and descriptive completeness*. The point is that they apply to different things.

1. *Semantic completeness* can be a property of an axiomatized system of logic or some of its fragments if the set of valid expressions of the system is recursively enumerable. This property can also be expressed by saying that the system has a full axiomatisation. As mentioned above, classical first-order logic enjoys this property, while IF logic does not.

2. *Deductive completeness* is related to a non-logical deductive system with an axiomatized system of logic. This property guarantees that, starting from a given non-logical system and applying given logical means, one can prove C or not C for every expression C of the language in question. Gödel's proof of the incompleteness of arithmetic concerns deductive completeness.

3. But there is another type of completeness that corresponds to the descriptive (and, according to Hintikka, the most important) role of logic in mathematics—Hintikka refers to this as *descriptive completeness*, which is actually *categoricity* in Dedekind's sense.[31] It applies to an axiomatized non-logical system and is related to the model-theoretic

---

[31] see Dedekind (1996).

approach. The descriptive completeness of the system exists if there is only one model, or—if there are more models—if one can prove that there is an isomorphism between them.

The impact of Gödel's result of the incompleteness of arithmetic has been widely discussed and Hintikka believes that it actually has limited importance. In Hintikka's view, for the foundation of mathematics it is the categoricity of a system that we should care most about, because it brings insight into the relationship between the theory and its instantiation i.e. the structures that satisfy it. Classical first-order logic, although semantically complete in the sense described above, cannot provide the categoricity of arithmetic. When we remain on the first-order level, there is no immunity to the emergence of non-standard models of theory, as follows from the Löwenheim–Skolem theorem. However, if one scarifies the semantic completeness of logic, one may be able to provide the desired isomorphism between models of arithmetic. This can be achieved by passing to second-order language or to set theory, as did Dedekind, Peano, and Frege, but while the categorical system of arithmetic can be obtained in this way, other problems emerge. Hintikka is a fierce defender of nominalism in mathematics and, according to him, it is essential in philosophy of mathematics to avoid a commitment to higher-order entities. On the other hand, set theory formalized with classical first-order logic in the usual way is a poor medium for model theory because the truth cannot be defined within it.[32] Hintikka would like to find a half-way point between the two: on the one side, to have a logic with the expressiveness of second-order logic in which the isomorphism between models of arithmetic is guaranteed; but on the other hand, to stay on the first-order level. According to Hintikka, IF logic is the right choice in every respect, so the failure of axiomatisability is not the issue. I should mention, however, that IF logic has a complete *disprove procedure*: invalid IF sentences can be recursively enumerated.

Later in this thesis I will compare Hintikka's approach with the dialogical approach that makes use of Constructive Type Theory. Both approaches are game-theoretic. However, as opposed to the model-theoretic approach, I will argue for an approach where games are oriented toward formal proofs. Hintikka was aware of a close link between GTS and dialogues but he saw the formal-proof orientation of the dialogical approach as its weakness. He argues the following (1973: 81):

> *In contrast to our games of seeking and finding, the games of Lorenzen and Stegmüller are 'dialogical games' which are played 'indoors' by means of verbal 'challenges' and 'responses'. […].*

---

[32] See Hintikka (1996a, 2006 etc)

[...]. *If one is merely interested in suitable technical problems in logic, there may not be much to choose between the two types of games. However, from a philosophical point of view, the difference seems to be absolutely crucial. Only considerations which pertain to 'games of exploring the world' can be hoped to throw any light on the role of our logical concepts in the meaningful use of language.*

However, the lack of a complete system of inference for IF logic conflicts Hintikka's claim that IF logic realises the logicist programme.

It is worth noting a few remarks made by Tennant (1998), concerning Hintikka's easy dismissal of a deductive aspect of logic. In his review of *Principles of Mathematics Revisited,* Tennant justly observes that the only axiomatisation Hintikka mentions therein is Hilbert-style axiomatisation, while *no mention is made of the much more natural and entirely rule-theoretic presentations given by Gentzen* (Tennant, 1998: 97). According to the author, this seems like an omission, given that Hintikka holds the *naturalness* of GTS.

Another of Tennant's criticisms concerns the lack of a clear link between the truth and falsity conditions in GTS and the usual requirement for inference-patterns in logic. This is even more important when the negation is taken into consideration, because the dual negation leaves different possibilities for defining validity: we would have one concept when Eloise has a winning strategy in every model and another when Abelard has no winning strategy in any model. This presents a serious ambiguity, according to Tennant (1998: 98). Moreover, the previously suggested complete disproof procedure suffers from the same problem.

Hintikka's sharp (and, one might say, scathing) response came in 2000, when he accused Tennant of a complete misunderstanding of game-theoretical semantics and of using *in his recent animadversions on IF logic (Tennant 1998) a disproportional amount of space and printer's ink to discuss rules for formal proofs of validity, which in the best of circumstances yield only indirect indications of the truth conditions that are the real life blood of logic* (2000: 135). No concrete response to Tennant's objections has been offered, to my knowledge.

I would like to make one more remark on this issue. A discussion about the advantages of a proof-theoretic approach over a model-theoretic one (or the other way around) is a big issue that I cannot address here. Questions concerning this are complex and demand an independent discussion. In this work I will undertake a more modest task: I will argue that neither the treatment of branching quantifiers nor that of anaphora really demands a model theory.

Before closing this section I would like to take a closer look at the way IF sentences prove validity. Clearly, the question is very important given that Hintikka proposes IF logic as the foundation of mathematics. I will cover this project of reducing most of the important mathematical theories to IF logic in more detail in section *1.7.*, but I have already pointed out that this reduction means that the question of the validity of mathematical sentences can be reduced to the question of the validity of IF sentences, where no quantification over higher-order entities is present. However, in the lack of a complete proof procedure the question of validity is far from being unproblematic.

To put it bluntly, in the lack of full axiomatization, proving the validity of IF sentences involves jumping to higher-order languages, which provides a reason to question the first-order character of IF logic.[33] Hintikka (1996a: 209–210) writes that:

> *My reduction of all mathematical theorizing to IF first-order logic needs to be viewed in a wider perspective. This perspective is provided by the distinction that I made in Chapter 1 between the descriptive and the deductive functions of logic and mathematics. If you review the reduction carried out in this chapter, you will see immediately that it concerns only the descriptive function of logic in mathematics. It concerns the question of what kind of logic is needed to capture and to master intellectually the structures (or classes of structures) mathematicians might be interested in. In contrast, my reduction does not mean that we can restrict to the first-order level the tools needed for the purpose of dealing with mathematical theories deductively. Indeed, I have apparently strayed unrealistically far from all questions concerning actual logical inferences and hence far from the deductive function of logic. For instance, in ordinary first-order logic, the truth of the conditional $(S_1 \rightarrow S_2)$ is precisely what is needed to move from the truth of S_1 to the truth of S_2. In contrast, the truth of $(S_1 \rightarrow S_2)$ (i.e., of $\sim S_1 \vee S_2$) in IF first-order logic is much more than what is needed to make this move in IF first-order logic.*

Indeed, the deductive role of logic is indispensable in mathematics, especially if someone proposes a realization of the logicist program. Let us focus a little more on the logical truth of IF sentences. In (1996a: 230) Hintikka reflects on the logical inference of one IF sentence from another. The truth of both sentences can be represented by second-order sentences expressing the existence of a winning strategy in a game played for each sentence, as we saw before. The question of whether the first sentence follows from the other will be then

---

[33] I will address this issue a bit later, in section *1.6.2.*

transformed into a question of the possibility of constructing a function g, expressing the winning strategy for the second sentence from a function f, that is, from the winning strategy for the first sentence. This can be formulated by a third-order sentence involving a functional that provides g as a function of f.[34] Hintikka (1996a: 231) goes on to say that when it comes to this deductive role of logic, it is reasonable to reach out to the constructive version of IF logic.

> As long as existence is taken in a standard sense, it does not follow that the verifier "has" a winning strategy in the sense of knowing what such a strategy is - or even being in a position to know what it is. In deduction, we are moving from the known truth of the premise or premises to the known truth of the conclusion. In order to handle such matters, it is mandatory at least to restrict the initial verifier's winning strategies to those that he or she can know. And this apparentl entails that these strategies can be represented by recursive (computable) functions.

> This means using the constructivistic logic outlined earlier in this chapter. In general, while it was not possible to find convincing reasons for adopting constructivistic standpoint in general in the foundations of logic and mathematics, there seems to be a great deal to be said for adopting it for the purposes of the deductive function of logic in mathematics.

I will come back to this issue later, where I will argue that Hintikka needs a constructive reading more that he seems to realize. First, I will argue that a constructivist reading is in fact required to assure the first-order character of IF logic. Second, I will challenge his claim in the previous quotation that the restriction to recursive functions actually yields such a constructivist reading of IF logic. At the end, I will question the very possibility of assuring a constructivist reading of IF logic in the lack of formal tools for inferences within that setting. But let me continue now with my presentation of IF logic and Game Theoretical Semantics.

---

[34] As Hintikka (1996a: 231) emphasis himself, this technique corresponds to Gödels's interpretation of conditionals.

I.5.8. The question of compositionality

Another important feature of IF logic is the failure of *compositionality*. The principle of compositionality requires that the meaning of a complex expression is determined by the meaning of its constituent parts and by its structure. In Tarski-type semantics the meaning is built up in this way: starting with atomic expressions, until the whole sentence is determined—the so called "inside-out" approach. In IF logic, on the other hand, this demand is impossible to meet: in IF logic an existential quantifier can be independent of some of the universal quantifiers within whose scope it lies and dependent on some others. The interpretation of such a sentence then violates the compositionality principle. Consequently, the game theoretical interpretation is much more natural for IF logic: it offers a different approach to meaning where we start off with the entire sentence until its constituent parts are reached.

Hintikka insists on the fact that the lack of compositionality is an advantage rather than a defect (Hintikka, 2002). That is, he relates compositionality to the *context-independence* of a sentence, while one should prefer semantics that provide the *context-dependent* meaning. Therefore, in Hintikka's opinion, IF logic *should not be* compositional.

Hodges proved in (Hodges, 1997a, b) that it is possible to create one variant of a compositional semantics for IF logic.[35] However, Hodges' semantics is not a Tarski-style semantics because in it we deal with *sets of assignments* and not with a single assignment, which is considered a price to pay. It is a set of assignments of elements to free variables that makes a sentence true or false in a given model and in this way the dynamic features offered by GTS are lost. Consequently, it is no wonder that Hintikka does not consider Hogdes' result to be a success. Indeed, in (Cameron, Hodges, 2001) it is proven that no strict Tarski-type semantics can be made for IF logic. In (Hintikka, 2002: 411), Hintikka states the following:

---

[35] A presentation of compositional semantics for IF logic can also be found in (Mann, Sandu, Sevenster, 2011).

This parallelism between syntactical and semantical rules is irretrievably lost in IF logic. The failure of compositionality which has thus been diagnosed, is closely related to the inadequacies of the traditional notion of scope. I have discussed the vagaries of this notion (with an eye also on its uses in linguistics) in an earlier paper (Hintikka [1997]) entitled 'No scope for scope?' Its punch line is: 'In linguistics, once a day with scope does not do it.' Wilfrid Hodges and his associates have tried to construct a compositional semantics for IF languages. In spite of their great ingenuity, which has led to important new insights into IF logics, the upshot is essentially an impossibility result, as far as my IF logic is concerned.

I will come back to the point Hintikka makes about *inadequacies of the traditional notion of scope*. In (Hintikka,1997a), he argues that the distinction should be made between the *priority* and *binding scope* of quantifiers. The priority scope concerns the relations of dependencies between quantifiers and other logical operators, while the binding scope indicates that the variable is bound by a certain quantifier. In traditional approaches there is some confusion over the two because they are both expressed by the same syntactic device. According to Hintikka, without a proper understanding of the two notions of scope we can never explain in a satisfactory manner some natural language phenomena such as anaphora. In (Hintikka, Kulas, 1983) it has been already argued that in order to provide a satisfactory semantics of natural language, compositionality has to be abandoned. In (Hintikka, 1996a: 108–109) there are some natural language counterexamples for compositionality—there the principle of compositionality is called *a good established dogma*. Hintikka (1996a: 110) concludes with the point that:

Conversely, to look at the bright side of things, the very existence of IF first-order logic is an eloquent proof that a rejection of compositionality is no obstacle to the formulation of a simple and powerful logic. Indeed, the best argument against compositionality as a general linguistic principle is the success of independence friendly logic in its different variants in the logical analysis of various important concepts and of their manifestations in natural languages. This is not the occasion to tell that success story, which is in fact still continuing. A cumulative evidence of different applications is in any case sufficiently impressive for me to rest my case against compositionality for the purposes of this book.

I will argue in what follows that the "outside-in" approach in game-theoretically oriented semantics actually has significant advantages over the traditional "inside-out" perspective. It will be illustrated later when dealing with anaphora. In addition, the distinction Hintikka makes between the *priority* and *binding scope* is indeed essential. However, I will compare and contrast GTS with the dialogical game theoretic approach to show that the *parallelism between syntactical and semantical rules* does not have to be lost along the way. Dialogical semantics is an alternative both to the model-theoretic and the proof-theoretic approach to meaning. Related to the Constructive Type Theory, it takes seriously meaning's dependence on the context in which the sentence is asserted and it has the tools to distinguish between the two notions of scope. But this can be done without neglecting the deductive role of logic. At the same time, a satisfactory interpretation can be provided for all of Hintikka's examples of problematic anaphoric expressions.

Moreover, I will argue that the "outside-in" or "top-down" approach, as it is sometimes called, is actually naturally related to pragmatist or *antirealist* approaches, rather then to the realist one. As Robert Brandom (2001: 13) explains:

> *Traditional term logics built up from below, offering first accounts of the meanings of the concepts associated with singular and general terms (in a nominalistic representational way: in terms of what they name or stand for), then of judgments constructed by relating those terms, and finally of proprieties of inferences relating those judgments. This order of explanation is still typical of contemporary representational approaches to semantics (paradigmatically Tarskian model-theoretic ones). There are, however, platonistic representational semantic theories that begin by assigning semantic interpretants (for instance, sets of possible worlds) to declarative sentences. Pragmatist semantic theories typically adopt a top-down approach because they start from the use of concepts, and what one does with concepts is apply them in judgment and action. Thus Kant takes the judgment to be the minimal unit of experience (and so of awareness in his discursive sense) because it is the first element in the traditional logical hierarchy that one can take responsibility for. (Naming is not a doing that makes one answerable to anything.) Frege starts with judgeable conceptual contents because that is what pragmatic force can attach to. And Wittgenstein's focus on use leads him to privilege sentences as bits of language the utterance of which can make a move in a language game. I take these to be three ways of making essentially the same pragmatist point about the priority of the propositional.*

I.5.9. Truth predicate

As per Tarski's well-known impossibility result of 1936, the truth predicate cannot be defined for a classical first-order language within the same language. Adding the truth predicate to a first-order language would produce paradoxes such as the famous *liar's paradox*.[36] In other words, first-order languages are not expressive enough to express their own truth. To be able to formulate this, a richer language is needed—a stronger meta-language. Hintikka labels the above result "Tarski's curse" (Hintikka, 1996a: 116).

The expressive power of IF logic exceeds that of classical first-order logic, enough to be able to define the truth predicate within the language itself. In Hintikka's words, IF language allows us to "exorcise Tarski's curse" (Hintikka, 1996a: 117); it is rich enough to be able to speak of itself. In (Hintikka, 1996a: 113) an adequate truth predicate is formulated for IF language using Gödel numbering.

When the truth predicate is added to IF language the liar's paradox does not occur because the liar sentence will be neither true nor false—its truth value will be undetermined. The possibility of formulating the truth predicate in IF language is actually due to the failure of the law of the excluded middle and to the fact that the negation used is not classical but stronger, dual negation. However, if we acknowledge that Tarki's impossibility result is restricted to first-order languages with classical negation, then it is not inconsistent with IF logic.

The fact that the truth can be defined in IF language is very important for the realist position Hintikka defends, especially if we bear in mind his model-theoretic perspective.[37] This is closely related to Game Theoretical Semantics, where the existence of a winning strategy for Eloise, which amounts to the truth in a model, includes the axiom of choice. When a sentence is true, Skolem functions exist for it, which renders it possible to express the truth in the language itself. But in the set theory formalised with classical first-order logic this is not possible, which makes set theory, according to the author, inadequate for a model theory in general. Hintikka (2008: 25) says that:

---

[36] See (Beall, Glanzberg, 2014).
[37] See the discussion in Hintikka (2008: 24)

This is a striking result in that it contradicts the widespread idea of axiomatic set theory as the natural medium of all model theory. This idea is simply wrong. For any halfway adequate model theory you need the notion of a truth, which just is not available in a set theory using traditional first-order logic. First-order axiomatic set theories are poor frameworks even for their own model theory. A fortiori, there are likely to be poor frameworks for any theory formulated in their terms.

Hintikka's claim about the definability of truth on the first-order level has been disputed. Marcus Pantsar presents arguments from De Rouilhan and Bozon (2006), that even if the truth predicate for IF language is adequate, it is only possible to show this in a richer language. Therefore, the hierarchy of languages cannot be avoided. Pantasar (2009: 164) argues that:

Finally, needless to say, first-order IF languages do not contain enough expressive power to carry out the sort of theorizing that Hintikka does in natural language. In this respect, De Rouilhan's and Bozon's criticism resembles my criticism of Field. It is all well to use axioms to define concepts in formal languages, but we should not forget that all this is initially done in our pre-formal languages. When we switch from pre formal to formal concepts, we cannot fool ourselves into thinking that the former never existed. Moreover, we cannot forget that they are the reason why the formal concepts are possible in the first place. Similarly, Hintikka's game-theoretic definition of truth for IF languages is adequate, but we cannot look at this result independently from all the background needed to establish it. De Rouilhan and Bozon argue that a first-order IF language itself is not able to express this, or any of the desired results around it. That is why Tarski's hierarchy exists also in IF. For us to be able to say that the definition of truth for IF is adequate, we need to have a richer language. The main problem for Hintikka (as well as for others that want to escape the Tarskian hierarchy) is that everything needs to be done in a single language. De Rouilhan and Bozon call this the monolingual speaker problem, and it follows us wherever we go with projects like Hintikka's.

Another argument of De Rouilhan and Bozon spelled out by Pantasar (2009: 165) is that:

...the model-theoretic concepts of logical truth, logical implication and logical equivalence are all definable in IF languages only in a very weak sense. Indeed, as they point out, here IF loses something from classical first-order languages where those concepts are fully definable.

However, as Pantasar notes, if the hierarchy of languages is still needed it is for different reasons than in the case of classical first-order logic. The possibility of expressing the truth in the language itself is still an important characteristic of IF logic. We will still discuss this point in relation to the axiom of choice in Chapter **III**. However, contrary to Hintikka's (2008: 6) claim, to take account of relations of dependencies and independencies between quantifiers on the first-order level and to be able to formulate the truth predicate in a particular language, IF logic is not *indispensable*. In Chapter **II** of this thesis it will be shown that the same can be carried out on the first-order level in the dialogical framework. Neither is a model theory indispensable, as I shall show.

I.5.10. Extended IF logic

Perhaps the most interesting element in IF logic is the so-called rule of *dual* negation. As presented above, this rule requires a change of roles for players in a given game. However, the problem is that IF logic with dual negation still proves insufficient to express all the important mathematical properties (among others, mathematical induction). Therefore, Hintikka proposes an extended version of IF logic, corresponding to a larger fragment of second-order logic, whose expressivity is fully satisfactory when it comes to mathematics. As Hintikka puts it, rather wittily:

> *At this point of my mystery story of the missing negation, I fully expect Lieutenant Colombo to scratch his head and say to me: "Yes, but there is still one question that bothers me. Why cannot we simply introduce a contradictory negation by a fiat? All we need to give a semantical meaning to such a negation, say ¬S, is to stipulate that ¬S is true just in case S is not true. What's wrong with such a procedure?*(Hintikka, 1996a: 147)

The extended version of IF logic is created by distinguishing two types of negation: *dual* or *strong* negation, denoted (~), and *contradictory* or *weak* negation (¬), that can occur only as a negation of an entire IF sentence. Semantically, this negation does not behave like other connectives, because there is no rule for it in terms of game theory. The semantic explanation is that $\neg\varphi$ is true if and only if φ is not true; in other words, if it is false or undetermined. The law of the excluded middle holds for contradictory negation. Contradictory negation in a sentence $\neg\varphi$ actually indicates that there is no winning strategy for the initial Verifier in a game played for φ.

The extended version of IF logic is even more expressive than regular IF logic. It corresponds to a broader fragment of second-order logic, the fragment that is usually denoted $\prod^1_1$. Contradictory negation fits a second-order sentence in the form $\forall x_1...\forall x_n \Psi$, where Ψ is a first-order formula. Together IF and extended IF logic are equivalent to the $\Sigma^1_1 U \prod^1_1$ part of second-order logic. This language is sufficient to express most important mathematical properties: mathematical induction, finite domains, well-ordering, and so on. Elementary number theory can be formulated with IF and extended IF logic and, moreover, its categoricity is guaranteed.

All Peano's axioms can be thus formulated, including the axiom-schema of mathematical induction. In its second-order version the axiom has the following form:

$$\forall X ((X(0) \wedge \forall y (X(y) \rightarrow X(y+1)) \rightarrow \forall y X(y),$$

which is an $\prod^1_1$ sentence.

However, introducing the contradictory negation changes a lot when it comes to the properties of IF logic. Pleasant properties like compactness, separation theorem, the Löwenheim-Skolem theorem, and the existence of a complete disproof procedure are lost.

I.5.11. Double negation law

Theorem:

If φ is an IF sentence then the following equivalence holds:

$$\varphi \equiv \sim\sim\varphi$$

Proof: This follows directly from the GTS rule of negation. Double negation means that players change their roles twice in a row in a game, so the strategy for winning

G ($\sim\sim\varphi$) is exactly the same as the strategy for winning G (φ).

I.5.12.De Morgan's laws

Theorem:

Let ψ be a subformula of an IF sentence φ. Then in IF logic the following equivalences hold:

a) $\sim(\psi_1 (\vee/ \forall x_1... \forall x_n) \psi_2) \equiv \sim\psi_1(\wedge/ \exists x_1... \exists x_n) \sim\psi_2$;

b) $\sim(\psi_1 (\wedge/\exists x_1... \exists x_n) \psi_2) \equiv \sim\psi_1(\vee/ \forall x_1... \forall x_n) \sim\psi_2$;

c) $\sim(\exists x/ \forall y_1... \forall y_n) \psi \equiv (\forall x/ \exists y_1... \exists y_n) \sim\psi$;

d) $\sim(\forall x/ \exists y_1... \exists y_n) \psi \equiv (\exists x/ \forall y_1... \forall y_n) \sim\psi$.

Proof: This follows from the GTS rules of connectives, quantifiers, and negation. In a), in the first part of the equivalence Abelard chooses which disjunct he will challenge (as the subformula is the negation). In the equivalent subformula the case is the same: by the rule of conjunction, it is Abelard's turn to choose between $\sim\psi_1$ and $\sim\psi_2$.

The proofs for b), c), and d) are similar.

Moreover, sentences a), b), c), and d) are true in the sense that the strategies for Abelard and Eloise are exactly the same in the games for both parts of each equivalence.

Given the double negation and De Morgan's laws, we can easily note how every IF sentence can be brought into the *negation normal form*. These laws allow us to push the negation sign until it reaches its atomic expression.

I.5.13. A remark on the equivalence

The preceding formulas are equivalent in the strongest possible way, meaning that two formulas are true in the same models and false in the same models. However, the failure of the law of the excluded middle yields two more weaker concepts of equivalence: when two formulas are true but not false in the same models, which is denoted ($\equiv_t$), and when two formulas are false but not true in the same models, which is denoted ($\equiv_f$). Hintikka actually understands equivalence as $\equiv_t$ equivalence, which can be illustrated by the quotation below (1996a: 65):

> *Are two sentences logically equivalent if and only if they are true in the same models (on the same interpretation), or are they to be so called if and only if they are true and false in the same models? In this work, I will consistently opt for the former alternative and for its analogies with other concepts of the metatheory of logic.*

This is not a big surprise if we keep in mind Hintikka's strong focus on truth—in his presentation of IF syntax only existential quantifiers and disjunctions can appear slashed, which are the decision positions for Eloise.[38] From Hintikka's model-theoretic perspective, material truth is the most important in mathematics, rather than *logical* truth[39]

It should be noted that the equivalence between IF logic and Σ^1_1 second-order logic is the $\equiv_t$ equivalence. IF sentences and the corresponding existential second-order sentences are always true in the same models, but not necessarily false in the same models. This is clear already from the fact that the law of the excluded middle holds for the second-order logic but fails in IF.

I.5.14. Compactness

Theorem:

> A set of IF sentences has a model if all its finite subsets have a model.
>
> *Proof:*[40] Let us denote a set of IF sentences α. Those sentences can be translated into existential second-order sentences using the skolemisation procedure described above. Obtained sentences have the form $\exists f_1 \ldots \exists f_k \forall x_1 \ldots \forall x_n (\varphi)$, where φ is a classical first-order formula. Let us denote this set β. Now in all sentences in β we omit the quantification over function symbols to get a new set γ, the set of classical first-order formulas. If every finite subset of γ has a model, then every finite subset of β and every finite subset of α has a model as well. The compactness theorem holds for classical first-order logic, so γ has a model if every

---

[38] Noted by Dechesne (2005: 32).
[39] See Hintikka (1996a: 66).
[40] Proved in Hintikka (1996a: 59).

finite subset of γ has a model. If γ has a model then α and β also have a model. Thus, if every finite subset of α has a model, then α has a model, which we set out to prove.

I.5.15. Löwenheim-Skolem theorem

Theorem:

> If a set of IF sentences has an infinite model then it has a model of any infinite cardinality.

Proof: The proof is similar to the previous one. We use sets α, β, and γ, as before. Sets of sentences α, β, and γ are satisfiable in the same models. Since the Löwenheim-Skolem theorem holds for classical first-order logic, if γ has an infinite model it has a model of any infinite cardinality. Then, if α has an infinite model it has a model of any infinite cardinality, which we set out to prove.

This is perhaps a good moment to discuss the so-called Skolem's paradox. The majority of logicians and philosophers recognise today that this is not the real paradox because it does not actually produce any contradiction in set theory.[41] However, Hintikka creates an argument against the classical first-order formalisation of set theory using this very paradox, claiming that it shows that set theory offers no clear account of basic concepts such as *countable* and *uncountable set.* Hintikka (1996: 202) says:

> *The Skolem paradox is less a paradox than an insight into the limitations of ordinary first-order logic. It is prompted by the Lowenheim-Skolem theorem which says that if a first-order sentence is satisfiable in an infinite model it is satisfied (true) in a countable model. What follows from this metatheorem is that first-order languages are not a satisfactory vehicle for discussing noncountable sets, for no first-order formula or axiom system can distinguish a noncountable set from countable ones. In the preceding chapter, it was indicated how these limitations apparently are manifested in (first-order) axiomatic set theory. Such applications have probably contributed to the air of paradox about the Lowenheim-Skolem theorem. This reputation is in reality thoroughly undeserved.*

---

[41] For a discussion of a paradox see Bays (2014).

Given that the Lövenheim-Skolem theorem holds in IF logic, it is interesting to see what response Hintikka offers to the so-called Skolem's paradox.

Let me first say a few words about the paradox itself. One could better describe it as a certain conflict or discrepancy between the Lövenheim-Skolem theorem, stating that if a first-order theory has infinite models, then it has models whose domains are only countable, and Cantor's theorem, stating that there are some uncountable sets. When we understand that Cantor's theorem is proven using the principles of a first-order theory that satisfies the Lövenheim-Skolem theorem, we face a puzzle. Notice that the Lövenheim-Skolem theorem is a meta-level theorem that says something about the first-order theory, while Cantor's theorem is an object-language statement. Later in this work I will argue for the advantage of a fully interpreted language, such as that of Constructive Type Theory, where the tension between the two statements cannot occur because everything essential on the meta-level has to be included in the object-language. On the other hand, McCarty and Tennant (1987) convincingly argue that in intuitionistic mathematics Skolem's paradox could not occur because even in a strong form of intuitionistic set theory the downward Lövenheim-Skolem theorem cannot hold.

A standard solution of a paradox is to say that it is actually a case of equivocation—Bays (2014) explains this in detail. To spell this out very briefly, quantifiers ranging over the model M, which is only countable, do not quantify over all the elements of all the members of M (some of which are uncountable sets). The model-theoretic interpretation of quantifiers is such that they range only over the domain of M and not over the whole universe of a set theory, which seems to be the ordinary English language interpretation. The confusion comes from the fact that the two interpretations do not coincide.[42]

As I have pointed out, Hintikka claims that, even though it is not much of a paradox, the puzzle nevertheless shows the shortcomings of the standard first-order formalization of set theory. He also claims that IF logic is a genuine first-order logic to which the Lövenheim-Skolem theorem applies. Let us examine his solution to the problem.

The idea is that any mathematical theory that has a finite second-order axiomatization can be formulated in extended IF logic, in which the Lövenheim-Skolem theorem does not hold. However, when it comes to proving whether a certain formula follows from a mathematical theory, the question concerns the validity of unextended IF formulas. According to Hintikka (1996: 203):

---

[42] For the solution of the paradox, see Resnik (1966), Kleene (1967) and Shapiro (1991).

This analysis of the situation can be thought of as a resolution of the Skolem "paradox". It shows that (and in what sense) mathematical theories can deal with uncountable structures even though their logic - at least the logic of theoremhood - can be handled in a logic that admits of the Lowenheim-Skolem theorem. In other words, we can now see how mathematical theories can deal not only with sets but even with uncountable sets, and yet mathematical reasoning is essentially combinatorial.

As pointed out by many mathematicians and philosophers, including Hintikka himself, Skolem's paradox is not a real paradox but rather a philosophical puzzle. In my opinion, the standard solution is quite a good response and, contrary to Hintikka's claim, it does not produce confusion about the set-theoretical take on sets. On the other hand, Hintikka's solution actually points to some weaknesses in his use of IF logic in mathematical theorizing: mathematical theories can be formulated only in extended IF logic, which does not enjoy many of the most pleasant logical properties, as I explained in section **1.5.10.** Among other problems, the truth predicate cannot be formulated in extended IF language. Questions of "theoremhood" should be carried out through questions of the validity of unextended IF formulas, but following that road requires a jump to higher-order logic, as explained above. This fact gives rise to serious questions concerning the order of IF logic, which I will discus below.

I.5.16. Separation theorem

Theorem:

Let φ and ψ be two contrary IF sentences[43] of vocabularies σ and σ', respectively. Then there is a classical first-order formula τ of vocabulary $\sigma \cap \sigma'$, such that $\varphi \equiv_t \tau$ and the sentences ψ and τ are contrary sentences.

Proof: The method is the same as in the previous proof. With the skolemisation procedure we get the existential second-order sentences φ' and ψ'. By omitting the quantification over function symbols in φ' and ψ' we get the classical first-order sentences φ'' and ψ''. The separation theorem holds for classical first-order logic, so there is a first-order sentence τ such that $\varphi'' \equiv_t \tau$ and the sentences ψ'' and τ are contrary

---

[43] Two contrary sentences have no model in common.

sentences. Sentences φ, φ', and φ" are true in the same models as well as ψ, ψ', and ψ". So τ is a first-order sentence such that φ ≡$_t$ τ and ψ and τ have no model in common.

I.6. Some problems with IF logic

I will move on to examine some of the problems with IF logic. The first is the so-called *signalling problem* that was first observed by Hogdes (1997b).

I.6.1. Signalling

The big novelty in IF logic is the slash sign that introduces imperfect information in a game played for a given IF sentence. The slash sign indicates that some information is hidden from one of the players in the game. Through the phenomenon of signalling it is possible for the information that is supposed to be hidden from the player to become available in the game via another variable.

Example:

Let the IF sentence φ be ∀x(∃y/∀x) (x = y) and the model $\mathbb{M}$ = {0, 1}. Neither Eloise nor Abelard has the winning strategy—the truth value of the expression is undetermined. But if we introduce a vacuous quantifier the situation will change. Note the following expression:

φ': ∀x∃z(∃y/∀x) (x = y).

In the game G (φ') it is Abelard who first makes his choice for variable x. Then Eloise makes her choice for z. In order to win the game, Eloise has to choose a value for z that is equal to the value Abelard has already chosen for x. Then it is Eloise's turn to move again. She has to choose the value of y, not knowing the value of x. But this time it is not a problem because she knows the value she has already chosen for z, which equals x, so she can choose the same value for y and win the game. Eloise has her winning strategy in the game for φ' due to the introduction of the vacuous quantifier ∃z. Formula φ' is now true in $\mathbb{M}$.[44]

The introduction of a vacuous quantifier cannot change the truth value of a classical first-order sentence. Games for classical first-order sentences are games with perfect information, so no signalling phenomenon is possible.

To prevent the signalling problem Hintikka introduces a convention: existential quantifiers and disjuntions are always independent of previous existential quantifiers in a sentence. In other words, Eloise forgets all her previous moves. Hintikka (1996a: 63) explains this as follows:

> *The small extra specification that is needed is that moves connected with existential quantifiers are always independent of earlier moves with existential quantifiers. This assumption was tacitly made in the second-order translation of IF first-order sentences explained earlier in this chapter. The reason for this provision is that otherwise "forbidden" dependencies of existential quantifiers on universal quantifiers could be created through the mediation of intervening existential quantifiers.*

This restriction works nicely in the previous example but unfortunately the problem does not end there. Janssen (2002) shows many other examples of signalling and raises new questions related to this non-desirable phenomenon. Janssen and Dechesne (2006) notice that the restriction made by Hintikka does not work well in some other cases where the signalling is actually *needed*. That can be illustrated by the following valid classical first-order formula:

$$\forall x \exists y \, (y = x \wedge (y = 1 \vee y \neq 1)).$$

It is easy to note that, with the restrictions introduced by Hintikka, Eloise does not have a wining strategy. If she forgets her previous moves then she must make her choice of a disjunct without knowing her previous choice for y, so she cannot win the game. The signalling phenomenon seems to be an essential part of GTS. Jansen and Dechesne conclude their article with a pessimistic observation:

> *Signalling is a tricky business. It disturbs several extrapolations from classical logic (change of bound variables, prenex normal from), and the interaction of signalling and implicit independence causes that Hintikka's IF is not a conservative extension of predicate logic.*

---

[44] The example is used first in Hodges (1997b).

Dechesne (2005: 104) argues that IF logic rests the conservative extension of classical first-order logic with one restriction: a classical first-order formula must be in its *regular form* as defined in *definition 10*, which ensures that there is no nested quantification over the same variable in the formula and that the same variable must not have both free and bound appearances in the formula. Janssen and Dechesne, (2006) are more pessimistic about this solution and suggest that not all examples of signalling can be prevented in such a way.

To end this section, let me quote an observation from Mann, Sandu, and Sevenster (2011: 74):

> ...*one can think of the semantic game for any IF sentence as a generalized signalling problem in which each existential quantifier corresponds to an agent who can both send and receive signals, while the universal quantifiers correspond to states of affairs beyond the agents' control.*

I.6.2. Is IF logic really a first-order logic?

According to Hintikka, the biggest advantage of IF logic is the possibility of expressing important mathematical properties on the first-order level. This claim has been challenged by authors such as Feferman, Väänänen, De Rouilhan, and Wolenski. The basic question is whether IF logic is really a first-order logic. Feferman (2006) claims that even though IF logic is a first-order logic syntactically, its semantics does involve higher-order entities. Hintikka's response is simple: the only criterion of whether a logic is first- or higher-order is the entities that the variables range over. Tulenheimo (2009) formulates this criterion in the following way:

> ...*a logic is of first order if any* play *of a semantic game associated with a formula of this logic only involves (in addition to choices interpreting conjunctions and disjunctions) choices of individuals, as opposed to choices of higher-order entities.*

This traditional view on the distinction, which relies on Quine's motto *to be is to be a value of a bound variable*, is called into question by Feferman and Väänänen. According to Feferman, the descriptive role of logic Hintikka favours so much in mathematics turns out to be insufficient. In mathematical reasoning, the deductive role of logic inevitably comes into play. When proving that a formula is a theorem of a system, we are dealing with the question of whether that formula follows deductively from the axioms of arithmetic. Let us observe the axioms of arithmetic that can be translated into extended IF language. Let us call such a sentence φ*. φ* has a form of a contradictory negation of a sentence φ of unextended IF language. The question is whether a sentence ψ is a consequence of φ*. This can be formulated as the question of whether the implication φ* → ψ is valid, which is equivalent to the disjunction ¬ φ* ∨ ψ (a formula of unextended IF logic). According to Hintikka, this is a way to reduce practically all of mathematics, or its greater part, to the question of validity in IF logic. But note that we are no longer speaking about the *satisfaction* of the sentence ¬ φ* ∨ ψ in a model; we speak now of its *validity*, i.e., of the truth in all models. Väänänen (2001) argues that when it comes to the validity of such formulas, we are committed to the entirety of second-order logic.

Determining the validity of an IF-logic sentence is to show that there is a winning strategy for Eloise in the game played for that sentence *in any model*. Tero Tulenheimo (2009) gives the example of a sentence ¬∀x∃y∀z(∃w/∀x) R(x, y, z, w), and asks whether we can avoid commitment to higher-order entities when we talk about its validity. If such a sentence is valid, it means that in each model there are functions f: M → M and g: M → M, such that the condition R (a, f (a), b, g (b)) is satisfied for the individuals a and b. To prove validity of this sentence we must assume that we have some understanding of *all functions* and *all subsets*, which is certainly a higher-order idea.

IF logic is necessarily oriented toward a model theory because it does not have a complete axiom system. As a consequence, we cannot really *do* mathematics in IF logic—and Hintikka is aware of this fact. Our mathematical practice has to remain unchanged and Hintikka's aim is purely foundational.[45] The idea is that, by expressing the most important mathematical concepts by IF logic, we get something important from the foundational point of view—we avoid the commitment to higher-order entities. I will discuss Hintikka's foundational intentions in more detail in the following chapter. Hintikka states in (1996a: 205-206) that:

> *What I have argued is that most (and perhaps all) of mathematics can in principle be done by means of one and the same logic, IF first-order logic. Moreover, that logic is a genuine article and not a disguised version of set theory because it is first-order logic and therefore free of the philosophical problems that have beset set theory or type theory. It has the same - and a better - claim to the title of logic as ordinary first - order logic. Needless to say, I am advocating this purely logical nature*

---

[45] Compare Hintikka (1996a: 206).

of mathematical theorizing as a rational reconstruction calculated to solve philosophical and other theoretical problems. As to practicalities, presumably the most intuitive way of mathematical theorizing would be to carry it out on a second-order level, or maybe with a simple type theory as its logic; in other words, not unlike what is done in general topology. Only when theoretical problems that have to do with the status of higher-order entities that have to do arise will there be a reason to resort to the kind of reduction to IF first-order logic described in this chapter.

I think that we should indeed embrace Feferman's view of the importance of inferential aspects of logic in mathematics. However, I think that, under some assumptions, the first-order character of IF logic can be defended. I will contrast Feferman's challenge with Goran Sundholm's analysis of IF logic to show that there are nevertheless some good reasons to consider it a genuine first-order logic. In the following chapters of this book I will however show that Hintikka's cherished goal of staying on the first-order level can be obtained, but without sacrificing the deductive role of logic in mathematics. Moreover, this can be done in a game-theory framework—but it will demand a dialogical approach to logic.

Let us look more closely at the question of the order of IF logic. Similar questions had been raised earlier, concerning the ontological commitment in dealing with branching quantifiers. Hand (1993) defended the standing that, if considered under the GTS interpretation, branching quantifiers demand no more ontological commitment than ordinary first-order quantifiers.[46] He argues against Quine and Patton's arguments in favour of the second-order reading of branching quantifiers. Patton's arguments seem more interesting from my perspective and so I think it is worth briefly mentioning some of them. The first is that a strategy in GTS is *a function*, that is, a second-order entity, and consequently, semantics employing such a notion make us committed to second-order entities. The argument is not so sharp and Hand (1993: 428) does not have much trouble dismissing it: in that case, one would have to say that under the GTS interpretation even classical first-order logic is second-order, because the truth definition for a classical first–order sentence involves a choice function - which would be a strange consequence indeed.

Second, Patton's argument for the second-order reading of branching quantifiers is more interesting. Patton (1991: 216) says that branching quantifiers should not in fact be considered quantifiers at all, because a necessary condition for

---

[46] Hand is employing the GTS interpretation of branching quantifiers and he even speaks of games with imperfect information. Curiously, he does not mention Independence Friendly Logic in this paper.

being a quantifier is having instances. Let us look once more at the typical example of branching quantifiers I presented previously:

$$\left.\begin{array}{l} \forall z \exists u \\ \\ \\ \forall x \exists y \end{array}\right\} \quad S(x,y,z,u)$$

There is no instance of such a sentence because what ever one could offer as its instance would have to be a linear sentence and any such a linear sentence would rather be rather an instance of a linear universal sentence and not of the given branching sentence.[47] As his answer to this objection, Hand proposes a re–evaluation of the concept of quantifier, made for classical first-order logic. According to him, we should extend the concept of quantifier so that it can include branching quantifiers. I will show later in this work that Patton's argument can be dismissed without extending the ordinary concept of quantifiers. The dialogical interpretation of branching quantifiers shows that these are indeed first-order quantifiers and they do indeed have instances. This is one more reason to favour dialogical semantics over GTS when it comes to the interpretation of branching quantifiers.

Ekland and Kolak (2002) remark that Hand's defence of the first-order character of branching quantifiers is strictly related to its GTS interpretation. They try to avoid this by defending IF logic on its own, even under some other semantic interpretations. However, their defence relies upon a certain conception of logic (however weak it might be) and on a certain claim about natural language—a claim they share with Hintikka. Their point is that IF logic should be considered first-order logic on the basis that this logic captures the fragment of the natural language that is usually considered to be first-order. Moreover, they follow Hintikka in his claim that IF logic can capture this first-order fragment while classical first-order logic cannot. They provided examples of natural language sentences that are considered first-order but cannot have a satisfactory interpretation in classical first-order logic; namely, sentences containing anaphoric expressions. I will elaborate on Hintikka's take on anaphora later. Finally, Ekland and Kolak arrive at a sort of

---

[47] See Hand (1993: 429).

relativisation of the very distinction between a first- and second-order logic. Indeed, the question of ontological commitment is not easy and the answer is not self-evident.

Another sceptical argument about the first-order character of IF logic is made by Cook and Shapiro (1998: 314–315) in their review of *The Principles of Mathematics Revisited*. They discuss Hintikka's claim that IF logic has the strongest case for being first-order logic through the simple fact that in it we avoid a quantification over functions and other second-order entities. The authors point out that, even if we accept that, there is still a problem with extended IF sentences. In the interpretation of these sentences there is rather an explicit quantification over functions. The contradictory negation does not have the same treatment in GTS as other connectives do—there is no game rule for it. A sentence that begins with the contradictory negation actually explicitly states the lack of a winning strategy on the part of the Verifier, i.e., it states the non-existence of a function. Therefore, there is no implicit but rather an explicit reference to a higher-order entity.

In my opinion, there is a very effective defence of the first-order character of IF logic against all previously mentioned challenges, formulated by Goran Sundholm (2013). Sundholm questions the very notion of a function in mathematics, stating that we should be careful when interpreting functions, sets, and other second-order notions. Sundholm makes use of Martin-Löf's Constructive Type Theory[48] to show that the functions the quantifiers range over in the Skolem form of IF sentences actually belong to the lowest level in the type-hierarchy. In CTT it is visible that in the Skolem form of IF sentences there are quantifications over the *function sets* and not over the *function types*. But *function sets* belong to the lowest level in the hierarchy because they are predicative and generated from below—as opposed to *function types* where this is not the case. A set in Constructive Type Theory is interpreted in terms of its canonical elements and of the way the set is composed out of those elements. As the author points out, the constructive conception of set remains the same whether the law of the excluded middle is included or not. In such a framework, a set is not interchangeable with its characteristic function. Therefore, if IF logic is considered within the Constructive Type Theory framework no quantification over function types is needed and, thus, there is no reference to higher-order entities. Hintikka can keep his ground. However, for this interpretation a particular take on the notions of set and function is needed; I will explore later in this book whether Hintikka can adopt this interpretation after all.

---

[48] With an important remark ...*that the considerations of this note are entirely neutral with respect to classical versus constructive logic because we can easily add the low of excluded middle to the type theory in question and the point will still stand.* (Sundholm, 2013).

I.7. Logicism revitalised

In this section I want to get a clearer picture of Hintikka's take on the role of IF logic in general and in the project of the foundation of mathematics in particular. With this I will conclude my presentation of IF logic and in the following chapter I will challenge some of Hinntika's points. Let us begin by following Hintikka's own line of thought.

According to Hintikka, IF logic has many uses. Among others, it is the most appropriate logic for the study of natural language. Hintikka considers it to be the most basic logic, over the classical first-order logic, because it allows for expressing relations between quantifiers that are not otherwise expressible on the first-order level. Hintikka (2002) labels IF logic *hyperclassical logic*. His point is that the fragment of natural language we traditionally associate with classical first-order logic is actually captured with IF language. The argument for this claim is that there are some sentences of natural language that cannot be captured by classical first-order formalisation yet still have an expression in IF language. As previously mentioned, those are natural language sentences involving branching quantification and anaphoric expressions. I will discuss this claim a little later.

In mathematical theorizing IF logic can play another important role. The fundamental question is that of the appropriate logic for mathematics. According to Hintikka, to answer this question we need first to reflect on the role of logic in mathematics. We have already underlined the three roles of logic distinguished by Hintikka: deductive, descriptive, and as a medium for axiomatic set theory. At first glance, classical first-order logic seems to be the best candidate for each of these functions because of the desirable logical properties it displays—such as completeness, compactness, the theorem of separation, the theorem of interpolation, Beth's theorem, and so on. Hovever, despite these advantages, classical first-order logic actually suffers from a range of problems in its application to mathematics. One of these is that not all mathematical concepts can be expressed in a first-order language. First-order languages are not expressive enough for concepts such as cardinality, well-ordering, infinity, mathematical induction, and so on. Another important problem, according to Hintikka, is the "curse of Tarski"—the impossibility of defining the concept of truth within the same language.

Consequently, we cross over to higher-order languages or to set theory. But there we find ourselves trapped into a new range of problems. In addition to the paradoxes of set theory, there is a problem of ontological commitment to sets and the issue of quantification over infinite domains. Hence, it would be very desirable for variables to range over individuals only. Furthermore, Hintikka argues that besides the issue of ontological commitment, standard set theory does not have a clear account of its intended models. Therefore, a search for stronger assumptions is not guided by questions of truth and falsity but rather by some vague intuitions.[49] As discussed in section *I.5.14.*, according to Hintikka, set theory is not a good candidate for the medium of any model theory. When it is formalised with classical first-order logic in the usual way, the problem of the definition of truth becomes essential. The impossibility of defining truth for a classical first-order language in the language itself is one of Hintikka's key arguments against set theory. Recall that the categoricity of mathematical theories is the crucial aim from the model-theoretic perspective Hintikka adopts. However, Tarski's result applies only to compositional formal languages. Hintikka refers to compositionality as *a good established dogma*. A non-compositional semantics allows formulating the truth predicate in IF first-order language.

IF logic provides the categoricity of arithmetic and is sufficiently expressive. Therefore, all important concepts of classical mathematics can be expressed in it. IF logic ought to have the advantages of both first and second-order logic, and to eliminate their problems. In addition, Hintikka believes that it allows for fulfilling the dream of realists when it comes to the foundations of mathematics.

I want to reflect a little more on Hintikka's realist position because it is quite specific. However, I will not be able to enter very deeply into a complicated discussion of mathematical realism—it is a large topic and demands separate consideration. Let me just say that, according to a very general and very simplified view on this matter, realism is a position holding that the truth of a proposition is independent of our knowledge or our ability to prove it. The meaning of the proposition is then grasped via its truth conditions. On the contrary, in the constructivist approach the meaning is not related to an objective concept of truth, independent of the human mind. The meaning is rather obtained by specifying the conditions for the assertability of a proposition and by providing the evidence that those conditions are fulfilled—the evidence being the actual construction of a proof for it. As pointed out in Rahman, Clebourn, and Jovanović (forthcoming):

> *Intuitionists, as pointed out by D. Prawitz (2012, p. 47) "avoid the term truth and reject the idea that intuitionism could replace 'p is true' with 'there exists a proof of p' understood in a realistic vein." Indeed, the existence of a proof, as pointed out by Prawitz in the same text, is to be "understood epistemically as the actual experience of the construction intended by the proposition, not as the existence of an ontological fact". More generally, from the intuitionistic point of view proof-theory*

---

[49] See Hintikka (1996b: 206).

provides the means for the development of an epistemic approach to meaning rooted in assertions (rather than propositions).

From another perspective, the difference between a realist and a constructivist has been described by Sundholm (2014: 4) as:

> *As so often in conflicts between Platonists and constructivists we have an instance of Fichte's (1797) dispute regarding epistemic priority between "Dogmatists" (that is, Realists or Platonists) and Idealists. The Platonist realist reduces the rightness of the epistemic act of knowledge to an object in the ontology, whereas for the constructivist the act is sui generis and yields the object of knowledge.*

Recall that the truth of a sentence in GTS amounts to the existence of a winning strategy for Eloise in a game played for that sentence. The existence of such a strategy is a fact concerning the structure of a model in respect to which the game is played. In a realist sense, it is a fact that is independent of the actual knowledge of a strategy function. As Ranta (1988: 383) explains:

> *The question is then weather we can introduce all games by stating their strategy prescriptions. For first - order games, the answer is affirmative. We can use a canonical notation for strategies, which enables us to read from a strategy term what game it is a strategy in. Conversely, given a first - order game, we can express in the canonical notation the form of a winning strategy of Myself in it. We can thus "bracket" the semantical games and say that a game is the set of winning strategies of Myself in it, as follows: given what canonical winning strategies of Myself in atomic games should be like, canonical winning strategies in complex games are defined in terms of them. A winning strategy in general, then, is an entity whose existence guarantees the existence of a canonical winning strategy. In constructive semantics this demand of guarantee is the strict demand that a non - canonical strategy be computable to a canonical one in order to count as a strategy.*

In Chapter 10 of Hintikka (1996a), we find a constructivist version of IF logic[50] in which choice functions are restricted to recursive functions or even to functions *known* to a player.[51] According to Hintikka, the former restriction yields

---

[50] Hintikka mentioned the constructive version of GTS much earlier, see Hintikka (1979: 21, 73).

[51] However difficult it might be to define the later restriction.

the constructivist approach while the intuitionist approach comes from the latter.[52] In constructivist IF logic it is required that all functions have to be *computable* or, in game terminology, we might say that strategies have to be *playable*.

This proposed restriction, which should yield a constructive version of IF logic, is discussed by Ranta (1988). He notes (1998: 387) that *this restriction somehow yields intuitionistic logic*. In my opinion, there are some good arguments showing that this is not the case. First, the "somehow" in the previous quote is not as innocent as it might seem. It seems that something more is needed to make the reading of GTS acceptable to a constructivist than just a restriction to recursive functions. The arguments are listed below.

- First, as Ranta points out, a constructivist cannot accept the GTS rule for negation as the change of roles of the players in a game. This rule yields $\neg\neg (A \vee \neg A)$ equivalent with $(A \vee \neg A)$, because the winning strategy for the former can be easily transformed into the winning strategy for the latter. However, the first expression is constructively provable while the other is not. Instead, in the constructivist reading $\neg A$ should be interpreted as $A \rightarrow \bot$.[53]

- Another important remark made by Ranta (1988: 387) is that in Hintikka's GTS, *language games are entirely theoretical entities with no connection with language use*. That is, in actual language-use sentences are *assertions, questions, commands*, and so on. In the constructivist approach the *force* they have in a conversation has to be accounted for. Therefore, when analysing a sentence two moments should be separated: a proposition and the force the sentence actually has in the argumentation.[54]

- The third point concerns the axiom of choice. In Chapter *III* of this book I will examine Hintikka's standing on the axiom. I will argue that to obtain a constructive reading of the axiom it is actually essential to have an *intensional* take on the choice function postulated in it. What really matters is the way the function is defined.

- Let me consider the last argument against Hintikka's plea that the restriction to recursive functions yields a constructive version of IF logic.

---

[52] Hintikka (2001) makes a distinction between *constructivism* and *intuitionism*.
[53] In the dialogical framework the negation is interpreted as $A \rightarrow \bot$ and with the particle rule for implication it amounts in a way to the change of roles of the players in a game.
[54] This requirement is also met in the dialogical framework.

It is supported by arguments provided by Skolem (1955), Heyting (1962), and Bishop (1967). The arguments are very well presented by Coquand (2014), where he shows that there is a conceptual gap between the notion of constructivity and the notion of recursivity. He concludes the article with the point that the notion of function in constructive mathematics cannot be defined in terms of recursivity. The main insight from the papers of Skolem, Heyting, and Bishop is the following.[55] Let us consider the definition of recursive function: if g $(u_1...u_n, x)$ is a recursive function such that for all $u_1...u_n$ there exists x such that g $(u_1...u_n, x) = 0$, then the function f which for $u_1...u_n$ provides the least x such that $g(u_1...u_n, x) = 0$ is a recursive function. The problem is the existential quantifier in the previous definition—it has to be understood constructively in order to obtain a constructive reading. But then the definition of recursivity already presupposes the calculability of a function. On the other hand, if the existential quantifier is not understood constructively the link between the recursivity and the computability of a function is lost.[56] Bishop (1967) suggested that functions in constructive mathematics should be understood in terms of "rules" and not in terms of recursivity. Bishop's suggestion supports my argument above: the reading of the axiom of choice in GTS is constructive if the choice function is defined *intensionally*. The recursivity of a function is not relevant for the constructive reading of the axiom.

To conclude this discussion, I want to make one more clarification concerning the very notion of constructive mathematics. The observation made by Skolem, Bishop, and Heyting that the notion of recursion cannot serve the analysis of constructive functions leaves open the possibility that the class of recursive and the class of constructive functions coincide.[57] I would like to make two remarks on this point. The first is that even if the class of constructive functions does extensionally coincide with the class of recursive functions, Hintikka cannot make use of that when it comes to IF logic. As I said before, there is a gap between atomic and complex formulas in IF logic such that atomic formulas are always determined in a model and so the law of the excluded middle holds for them. Thus, a restriction to recursive functions would never yield a constructive IF logic. Second, and somewhat more general, is a point concerning the very understanding

---

[55] I follow here Coquand's interpretation (2014).
[56] The argument is presented by Heyting (1962).
[57] I thank for this remark to Dr Mark Van Atten who made it during the defence of my thesis. Moreover, he observed that if Hintikka adds Church's Thesis to the argument that constructive functions coincide with recursive ones, many constructivists might agree with him. Indeed Hintikka does make the appeal to Church's Thesis in another context.

of constructivism. There are several streams of constructivism, but I want to point to one that I follow in this work alone. In the text mentioned above, Coquand (2014) argues (following Bishop) that in constructive mathematics it is conceptually more satisfactory to introduce functions without reference to any specific algorithm, and many authors today share his view on the matter.[58] In this spirit Tait (2006: 213) writes:

> *Let me say straight off that there are two distinct ideas: one is construction and the other is computation. These have been confused in recent history, but really are distinct. "Constructive" means that the only witnesses of existential propositions one admits are ones that can be constructed, where of course this implies some background rules of construction. From the construction of an object, a means of computing it (in cases in which this idea makes sense) may or may not be found.*

Moreover, it is visible in the dialogical approach to logic that constructive functions should be presented in a game-theoretical framework as rules resulting from interactions rather then defined via notion of recursivity. In this way, constructivism is obtained without reference to any specific algorithm.

In Chapter *II* I will present the dialogical approach to logic linked with Constructive Type Theory and clarify the points mentioned above. In dialogical logic all requirements are met to make a constructivist reading possible in a game-theoretical approach. I will reflect on this point below.

However, Hintikka argues against a constructivist reading of GTS. He is in favour of embracing the whole of classical mathematics—curiously, by making use of a non-classical logic. Here is one of his arguments against constructivism:

> *Even though many constructivists emphasize the role of human thinking and human constructions in the "game" of mathematics, the main argument on their behalf virtually amounts to arguing that since robots have to be constructivists, we humans, too, have to be constructivists. For the only strategies that a digital automaton can be programmed to play in accordance with are the recursive ones. This is precisely what the proconstructivist argument sketched above has alleged to be the inevitable human predicament. The constructivists are in effect imputing to human beings those very limitations that characterize digital computers. It is thus the classical mathematicians who have more faith in human creativity than constructivists, contrary to the occasional claims of the latter.* (Hintikka, 1996a: 227)

---

[58] See Sundholm (2014).

As pointed out by Ranta (1988: 379), Hintikka contests constructivist proof-conditional semantics but his main argument is in fact erroneous: *that proof conditions cannot be given for branching quantifier sentences*. I will show later that it is not true in the light of Martin-Löf's Constructive Type Theory.

Let me reflect now more on the question of ontological commitment. Models of mathematical theories are structures made of certain elements. The structure of natural numbers is one such model for elementary number theory. I should give an outline of the ontological status of these primitive elements in Hintikka's theory. With regard to this issue, his realism goes hand in hand with nominalism, as he himself emphasises in (1996a: Ch 10, 2006b). In Hilbert's sense, individuals in a model are understood as individual extra-logical objects, directly accessible to human experience, like records on paper or a board. Curiously, according to Hintikka, Hilbert should be considered a realist—the formalism often associated with him is erroneous.[59] Hintikka's reading of Hilbert focus on his finitism and his *preference for combinatorial over set-theoretical reasoning in the foundations of mathematics* (1997b: 16). Combinatorial reasoning is introduced via epsilon-terms interpreted as choice functions, although Hilbert's weak point following this interpretation is that relations of dependencies are neglected in these choices. However, Hilbert's endeavor should be first and foremost understood as a justification of nominalism in mathematics—the primitive elements forming structures must be first-order individuals and not sets, because sets, having their own structures, would make a difference in a reformulation of axioms and in performing deductions from them.[60] It is disputable whether this is a good interpretation of Hilbert. It seems that Hilbert's purely proof-theoretical orientation in proving the consistency of mathematical theories is easily characterized as a less important technical tool for proving the existence of models for those theories and for dealing with them in a combinatorial manner.

In IF logic we reduce quantifications over higher-order entities to the combinatorial problem of *seeking and finding* individuals in a model. The word *combinatorial* is understood in a specific sense to refer to combinations of objects, not formulas. The advantage is that in IF logic there is an ontological commitment to first-order entities only. In a sense, this reduction is "the dream come true" of logicists, as Hintikka emphasises, in a modified and less ambitious version: it is a conceptual reduction, and not a translation of mathematics to logic. The idea is that classes of models of most mathematical theories can be grasped with IF logic and extended IF logic. Hintikka argues that this reduction is the real foundational reduction. All questions concerning the existence of higher-order entities can thus

---

[59] See Hintikka (1997b).
[60] See Hintikka (2006b).

be reduced to the issue of the truth or validity of IF sentences. Mathematics is considered to be research into the structures or models of a theory where structures are made up of individuals and not of abstract entities. Hintikka (1996a: 206) states:

> *My reduction of all mathematical problems to questions concerning the validity of sentences of an IF first-order language has both mathematical and philosophical significance. One kind of mathematical significance is that it shows that practically all mathematical problems are at bottom combinatorial rather than set-theoretical. This implies that the notion of truth applies in mathematical theories. If you look at set theory, especially in its familiar axiomatic dress, you have a theory whose intended models are not clearly understood, so that the choice of stronger assumptions seems not to be guided by questions of truth and falsity but by some vague "intuitions", or else by considerations of mathematical taste and expediency. In contrast, combinatorial problems are clear-cut. Either there exists a structure of a certain kind or else there does not exist one. Either your jigsaw puzzle or tiling task can be completed or else it cannot be. We have in such cases a razor-sharp characterization of the structures whose existence we are speaking of. This provides no reason whatsoever for dispensing with the notion of truth. The search for stronger deductive premises will be guided by one's combinatorial experience.*

In my opinion, there is an issue with the proposed logicist programme I mentionned above: IF logic has no underlying theory of inference. We cannot actually *do* mathematics in IF logic.

In Hintikka's own words, GTS is a semantics that overcomes the distinction between the verificationalist and vericonditionalist approach. The meaning is obtained, in Witgenstein's manner, through certain actions. Games for quantifiers amount to seeking and finding individuals in a model and the truth of a sentence amounts to a combinatorial problem of finding suitable *witness-individuals*. According to Hintikka, GTS, in its core, should already be acceptable for a constructivist. As shown previously, the very notion of truth in GTS yields a version of the axiom of choice. Hintikka thinks that GTS shows how a constructivist should embrace this classical principle.

In what follows I will make use of Martins-Löf's analysis of the axiom of choice to show that there is indeed a version of the axiom of choice that is perfectly acceptable for a constructivist, but it is a version of the axiom where the choice function is understood *intensionally*. However, to embrace the entirety of classical mathematics *extensionality* is required. To have both at the same time seems to amount to having one's cake and eating it too.

Moreover, I want to suggest more than that. In another passage against the constructivist approach Hintikka (1996a: 232) says:

> *First, in this way constructivistic ideas can be seen to play a major legitimate role in a mathematician's work in any case. They are needed to understand and to master the deductive task of logic in mathematics. At the same time, this role is different from what constructivistic philosophers like to think. For one thing, the role of constructivistic notions has nothing to do within the meaning of mathematical statements. Meaning is a matter of the descriptive function of logic in mathematics. It is a matter of sentence-model relationships, ultimately a matter of truth definitions. It is not a matter of deductive relationships between propositions.*

I will contest Hintikka's view from the last quotation in the following chapters using a constructivist approach, where the dialogical logic and Martin-Löf's Constructive Type Theory are linked together. I will show that the meaning can be explained in a satisfactory manner without sacrificing a complete proof system. To that goal I will examine two of Hintikka's favorite examples—the axiom of choice and anaphora. But let me first present Constructive Type Theory and the dialogical approach to logic.

PART II

Constructive type theory in a dialogical framework

In this chapter I will present Martin-Löf's Constructive Type Theory in the dialogical framework. The aim of this chapter is to provide some convenient tools that will later be used to question some of Hintikka's points. The presentation will not be exhaustive and I will not be able to discuss it in every detail, but I will explain as much as necessary for my purposes here.

II 1. Constructive Type Theory (CTT)

The father of type theory is Russell. Russell introduced the theory as a cure for paradoxes arising in set theory. The first version of type theory was called *simple type theory*, in opposition to the later *ramified type theory* from *Principa Mathematica*. Church made the standard formalization of simple type theory in terms of λ-calculus. An important contribution to further development was made by Curry and Howard, who used it in the theorization of computer programming. The first study of dependent type theory was done in 1960 by de Bruijn and his collaborators in connection with the AUTOMATH project. Those works inspired Martin-Löf to formulate CTT.[61]

The central idea of Martin-Löf's CTT is the *"propositions-as-types"* principle. According to this principle, propositions are identified with sets, types, or sets of their proof-objects. CTT has intuitionistic logic at its core and is fully predicative. There are several features of CTT that I would like to bring to the fore:

- In CTT a clear distinction is made between judgments (or assertions) and propositions expressed by those judgments. In other words, propositions are contents of judgments. Judgments are certain linguistic acts that cannot be considered independently of the *context* in which they are made. This important notion of context makes the CTT approach to meaning fundamentally pragmatic, as pointed out by Ranta (1994: 2).

---

[61] See Per Martin-Löf (1984). It was further developed by Sundholm (1986) and Ranta (1994).

This feature, among others, means that it is very appealing to make a link between CTT and the dialogical approach to logic, as I will show later.

• Classical logic can be developed in the CTT framework.[62]

• As opposed to a model-theoretic approach to meaning, where elements of a language are linked to the world (or interpreted) via some meta-level means, in CTT judgments and inferences are embedded at the object level and they account, by means of inferential rules, for the meaning of every expression, such that a fully interpreted language results. This carries out Frege's idea of a *Begriffsschrift*—which should provide a full and explicit formalization of mathematics.

• We saw in the previous chapter that GTS, as worked out by Hintikka, is oriented towards a model theory and used by Hintikka to defend a realist position. However, a game-theoretical approach does not have to be either of these: in what follows, I will present a dialogical game-theoretical approach that is not linked to a model theory and has a constructivist origin. More of the common ground for dialogical logic and CTT will be explained in the following sections.

II 2. The dialogical approach to logic

The dialogical approach to logic, in its contemporary form, was first proposed by Paul Lorenzen in the mid-20th century and later developed by Kuno Lorenz.[63] Most recent developments in this field have been made by Shahid Rahman and his colleagues.[64] Dialogical semantics is shown to be the most fruitful

---

[62] see Ranta (1994: chapter 2).

[63] Their most important papers about dialogical logic are published in Lorenzen and Lorenz (1978). See also Lorenz (2001).

[64] The first major development on the dialogical logic after work of Lorenzen and Lorenz was made by Rahman (1993) and since then by Rahman and Keiff (2005), Keiff (2009) and Rahman (2012). For the underlying metalogic see Clerbout (2013a,b). For textbook presentations see Kamlah and Lorenzen (1972, 1984), Lorenzen and Schwemmer (1975), Redmond and Fontaine (2011) and Rückert (2011a). For the key role of dialogic in regaining the link between dialectics and logic, see Rahman and Keff (2010). Keif

for a study of the validity of propositional and first-order formulas, both in their classical and intuitionistic versions,[65] but also as a powerful tool for the study of some non-classical logics. In the somewhat different, rhetorical tradition, dialogues are shown to be significant for a study of natural languages and for the theory of argumentation.[66]

The dialogical approach to logic is a game-theory semantic framework. A *dialogical game* is played by two players: the Proponent, who posits *the thesis* and defends it, and the Opponent, who tries to falsify it. The actions or moves of the two players can be understood as elements in the argumentation or certain speech acts, which are of two kinds: declarative utterances (posits) and interrogative utterances (requests). The actions are guided by fixed rules, which provide the *meaning* of expressions in such a way that the meaning is not independent of the very act of uttering the expressions. This is the pragmatic feature that dialogical logic shares with CTT and that makes their linkage a very natural one.

A dialogical game for a sentence φ starts off with φ posited as the thesis by the Proponent. Then the players make moves one after the other, according to the rules, and such a sequence of their moves is called *a play*. Dialogical game is then the set of all possible plays for φ. It can be provided a tree representation: the root is the thesis and every path starting with the root is a linear representation of a play.

The dialogical approach to meaning and GTS, both being game-theoretically oriented, apparently have a lot in common.[67] Both in dialogues and in the GTS approach the meaning is established in so-called "outside-in" direction, as opposed to Tarski-style "inside-out" semantics. The meaning of an expression is obtained in a game played for that expression, starting with the whole expression

---

(2004a,b, 2007) and Rahman (2009) studied Modal Dialogical Logic. Fiutek et al. (2010) studeid the dialogical approach to belief revision. Clerbout, Gorisse and Rahman (2011) studied Jain Logic in the dialogical framework. Popek (2012) developed a dialogical reconstruction of medieval *obligationes*. Rahman and Tulenheimo (2006) studied the links between GTS and Dialogical Logic. For other books see Redmond (2010) on fiction and dialogic, Fontaine (2013) on intentionality, fiction and dialogues and Magnier (2013) on dynamic epistemic logic and legal reasoning in a dialogical framework.

[65] Dialogical logic for first-order logic is presented in the Appendix.
[66] This approach is developed by Perelman. See Keiff (2009).
[67] The link between the two approaches was first presented in the paper of Esa Saarinen in 1978. Not much work was done in this field until 2006 when Rahman and Tulenheimo studied the relation between GTS and dialogues for the propositional and first-order logic in a systematic way.

and then reaching its component parts. However, dialogues have a more subtle apparatus for approaching the meaning on which I will elaborate later in this work.

As in GTS, in the dialogical approach the distinction is made between the *play level* (that is, winning of a local play) and the *strategy level* (the existence of a winning strategy). The notion of validity in formal dialogues and the notion of satisfaction in material dialogues amount to the existence of a winning strategy for the Proponent in a given dialogical game.

However, there are some important differences between the dialogical approach to logic and GTS:

- First, unlike in GTS, in the dialogical framework there are two kinds of rules: *particle rules*, which are the rules for logical constants, and *structural rules* which determine the general course of a dialogical game. This division allows for making a difference between the *local* and *global* meaning. The local meaning is player-independent.

- The dialogical framework is not model-theoretic. A formula is said to be valid if there is a winning strategy in a dialogical game played for the formula *independently of any model*. Recall that the notion of validity in GTS is understood as the existence of a winning strategy for Eloise in *every model*.

- In dialogues the distinction is made between *formal dialogues*, where the validity of a formula is determined, and *material dialogues* where the satisfaction of a sentence is at stake. In the former, turns are played independently of knowing the meaning of elementary sentences in the main thesis. In the latter, the game is relative to a model. However, the idea is still to avoid specifying the model explicitly. Instead, a list of additional hypotheses is introduced as the *initial concessions* of the Opponent. Those hypotheses then form the root of the dialogue, together with the thesis defended by the Proponent. As explained by Rahman and Tulenheimo (2006: 24-25):

> By contrast, the idea behind material dialogues is to avoid having an extra component to dialogues (such as a specification of a model); they are meant to do with the resources of dialogues designed for dealing with validity, and the idea is to 'approximate' a characterization of truth by adding a sufficient amount of additional hypotheses — taken to be initial

*concessions of Opponent — which will serve to specify a model by using
the resources of the object language only.*

Rahman and Tulenheimo (2006) established a connection between GTS
and the intitionistic material dialogues for first-order logic in a constructive way.
They showed precisely how to construct a winning strategy for the Proponent in
the dialogue D(φ) for a first-order sentence φ, starting with the winning strategy for
Eloise in the semantic game G(φ) and, conversely, how the winning strategy for the
Proponent in the material dialog D(φ) gives rise to the winning strategy for Eloise
in the semantic game G(φ).

II.3. The dialogical approach to CTT[68]

Let me move on to present recent developments in dialogical logic, where
a link has been made between dialogical logic and CTT.

As mentioned previously, the idea behind CTT is to render meta-logical
features explicit in object-language. The reader might remember Wittgenstein's
arguments against formal Tarski–style semantics,[69] which is one of Wittegenstein's
tenets that Hintikka (and othere model theoreticans) rejects. Moreover, the aim of
rendering explicit the rules of meaning-building is rooted in the dialogical
approach, where the meaning is constituted by certain interactions between players.
It is only natural to request that those interactions be rendered explicit in an object-
language. Indeed, one of the main insights of Lorenz's (1970: 74–79) interpretation
of the relation between the so-called *first* and *second* Wittgenstein is based on a
thorough criticism of the metalogical approach to meaning. As pointed out by
Kuno Lorenz, the heart of Wittgenstein's philosophy of language is the internal
relation between language and world. The internal relation is what language games
display while they constitute meaning: there is no way of grounding a logical
language outside of that language (recall the case of Neurath's sailor in his raft):

> *Also propositions of the metalanguage require the
> understanding of propositions, […] and thus cannot in a
> sensible way have this same understanding as their proper*

---

[68] For a detailed account see Rahman and Clerbout (2013, 2014). It will also appear in
Rahman, Clerbout, and Jovanović (forthcoming).
[69] For a discussion see (Rahman, Clerbout, Jovanović, forthcoming).

object. The thesis that a property of a propositional sentence must always be internal, therefore amounts to articulating the insight that in propositions about a propositional sentence this same propositional sentence does not express anymore a meaningful proposition, since in this case it is not the propositional sentence that it is asserted but something about it.[70]

Thus, if the original assertion (i.e., the proposition of the ground-level) should not be abrogated, then this same proposition should not be the object of a metaproposition, [...].[71]

While originally the semantics developed by the picture theory of language aimed at determining unambiguously the rules of "logical syntax" (i.e. the logical form of linguistic expressions) and thus to justify them [...] – now language use itself, without the mediation of theoretic constructions, merely via "language games", should be sufficient to introduce the talk about "meanings" in such a way that they supplement the syntactic rules for the use of ordinary language expressions (superficial grammar) with semantic rules that capture the understanding of these expressions (deep grammar).[72] (Lorenz, 1970: 109).

---

[70] Similar criticism has been raised by Sundholm (1997, 2001) who points out that the standard model-theoretic approaches to meaning turn semantics into a meta-mathematical formal object where syntax is linked to semantics by the assignation of truth values to uninterpreted strings of signs (formulae). Language does not any more *express content* but it is rather conceived as a system of signs that speaks *about* the world - provided a suitable metalogical link between signs and world has been fixed

[71] The quote is a translation of the following original text:

Auch Metaaussagen, so können wir zusammenfassen sind auf das Verständnis von Aussagen, [...] angewiesen, und können dieses Verständnis nicht sinnvoll zu ihrem Gegenstand machen. Die These, dass eine Eigenschaft eines Aussagesatzes stets intern sein muss, besagts daher nichts anderes, als die Artikulation der Einsicht, dass in Aussagen über einen Aussagesatz selbst nicht mehr der Ausdruck einer sinnvollen Aussage ist, nicht er wird behauptet, sondern etwas über ihn.

Wenn also die originale Behauptung, die Aussage der Grundstufe nicht ausser Kraft gesetzt werden soll, darf sie nicht zum Gegenstand einer Metaaussage gemacht werden, [...]. (Lorenz 1970: 75).

[72] The quote is Rahman's translation of the following original text:

Diente ursprünglich die mit der Abbildtheorie entworfene Semantik dazu, die Regeln der 'logischen Syntax', also die logische Form sprachlicher Ausdrücke, eindeutg zu bestimmen und damit zu rechtfertigen [...]-, so soll jetzt der

Hintikka (1996b) extends van Heijenoort's distinction between *language as the universal medium* and *language as a calculus*—the point is that the dialogical approach shares some tenets of both conceptions. Indeed, on the one hand, the dialogical approach shares with the universalists the view that we cannot place ourselves outside our language; while on the other hand, it shares with anti-universalists the view that we can develop a methodical reconstruction of a given complex linguistic practice out of the interaction of simple ones.

II. 3.1. Rules of formation of propositions

It is time now to presentat the dialogical rules for Constructive Type Theory. First, there are rules for the formation of propositions. These rules are local-semantic rules that together with particle rules provide meaning to the logical constants in game-theoretical terms. The formation rules are indispensable for meeting the request for a fully interpreted language. By those rules the players are allowed to ask explicitly during a game for the status of expressions. Otherwise, if an expression is well formed or not would have to be examined on the meta–level, which is the feature we want to avoid in this approach.

The formation rules actually allow Opponent to question both syntactic and semantic features of the thesis. Those rules bring the main requests of CTT to the fore: request that propositions are identified with sets and that those sets are constructible. So, if Proponent posits the thesis φ, Opponent can ask him about the formation of φ–then Proponent can defend his position by stating that φ is a proposition (written φ: prop) under condition that A is a set (A: set). Opponent will then concede that A is the set given that Proponent can demonstrate how A was constructed out of its elements.

When it comes to material dialogues, a rule concerning elementary sentences is indispensable. In principle, an elementary sentence can be attacked only by Opponent, through the formation rules. Here is the rule:

Sprachgebrauch selbst, ohne Vermittlung theoretischer Konstruktionen, allein auf dem Wege über die 'Sprachspiele', zur Einführung der Rede von 'Bedeutungen' hinreichen und die syntaktischen Regeln zur Verwendung gebrauchsprachlicher Ausdrücke (Oberflächengrammatik) mit semantischen, das Verständnis dieser Ausdrücke darstellenden Regeln (Tiefengrammatik), ergänzen. (Lorenz 1970: 109).

*O's elementary sentences cannot be challenged, however, **O** can challenge an elementary sentence (posited by **P**) iff she herself (the opponent) did not posit it before.*

The formation rules are the following:

Posit	Challenge [when different challenges are possible, the challenger chooses]	Defence
$X ! \Gamma : set$	$Y ?_{can} \Gamma$ or $Y ?_{gen} \Gamma$ or $Y ?_{eq} \Gamma$	$X ! a_1 : \Gamma$, $X ! a_2 : \Gamma$, ... (**X** gives the canonical elements of Γ) $X ! a_i : \Gamma \Rightarrow a_j : \Gamma$ (**X** provides a generation method) (**X** gives the equality rule for Γ)
$X ! \varphi \vee \psi : prop$	$Y ?_{F\vee 1}$ or $Y ?_{F\vee 2}$	$X ! \varphi : prop$ $X ! \psi : prop$
$X ! \varphi \wedge \psi : prop$	$Y ?_{F\wedge 1}$ or $Y ?_{F\wedge 2}$	$X ! \varphi : prop$ $X ! \psi : prop$
$X ! \varphi \rightarrow \psi : prop$	$Y ?_{F\rightarrow 1}$ or $Y ?_{F\rightarrow 2}$	$X ! \varphi : prop$ $X ! \psi : prop$
	$Y ?_{F\forall 1}$	$X ! A : set$

$X \,!\, (\forall x : A)\, \varphi(x) : \text{prop}$	or $Y \, ?_{F\forall 2}$	$X \,!\, \varphi(x) : \text{prop}$ $(x : A)$
$X \,!\, (\exists x : A)\, \varphi(x) : \text{prop}$	$Y \, ?_{F\exists 1}$ or $Y \, ?_{F\exists 2}$	$X \,!\, A : \text{set}$ $X \,!\, \varphi(x) : \text{prop}$ $(x : A)$
$X \,!\, B(k) : \text{prop}$ (for atomic B)	$Y \, ?_F$	X *sic* (n) (X indicates that Y posited it in move n)
$X \,!\, \bot : \text{prop}$	–	–

II. 3. 2. Rules of substitution

Players are allowed to ask for a substitution in the context $x_i : A_i$ either if a variable (or number of them) appears in a posit with a proviso (then the challenger posits an instantiation of the proviso) or in a formation-play. Here is the substitution-rule:

Posit	Challenge	Defence
$X \,!\, \pi(x_1, \ldots, x_n)\, (x_i : A_i)$	$Y \,!\, \tau_1 : A_1, \ldots, \tau_n : A_n$	$X \,!\, \pi(\tau_1 \ldots \tau_n)$

II. 3. 3. Particle rules

Recall that in CTT propositions are identified with sets of their *proof-objects*. In the dialogical context, propositions are rather understood as sets of *play-objects*.[73] Notation *p: φ*, which is the form of players' posits in a game, will be read as *p is a play-object for φ*. The meaning of those posits depends, in a game–theoretical manner, on the way how the play-objects are used in the game, that is, how those posits are challenged or defended. The form of play-objects depends on the structure of an expression φ and they can be obtained of other play–objects during the game, according to the particle rules. Particle rules are the following:

Posit	Challenge	Defence
X φ (where no play-object has been specified for φ)	**Y** ? play-object	**X** p : φ
X p : φ∨ψ	**Y** ?$_{prop}$	**X** φ∨ψ : prop
	Y ?[φ,ψ]	**X** L$^{\vee}$(p) : φ[74] Or **X** R$^{\vee}$(p) : ψ **[the defender has the choice]**

---

[73] Play-objects are actually different from proof-objects in CTT. It can be read from the treatment of the universal quantification and implication, which is different in the dialogical approach and in CTT. For discussion about it see (Rahman, Clerbout, 2014: 30) - as the authors point out there, the connection between dialogues and CTT is in the level of strategies rather then in the level of plays.

[74] L and R are the abbreviations for the Left and Right part of the expression.

	Y ?$_{prop}$	**X** φ∧ψ : prop
X p : φ∧ψ	**Y** ?[φ] Or **Y** ?[ψ] **[the challenger has the choice]**	**X** L^(p) : φ respectively **X** R^(p) : ψ
	Y ?$_{prop}$	**X** φ→ψ : prop
X p : φ→ψ	**Y** L→(p) : φ	**X** R→(p) : ψ
	Y ?$_{prop}$	**X** ¬φ : prop
X p : ¬φ	**Y** L⊥(p) : φ	**X** R⊥(p) : ⊥
	Y ?$_{prop}$	**X** (∃x : A)φ : prop
X p : (∃x : A)φ	**Y** ?$_L$ Or **Y** ?$_R$ **[the challenger has the choice]**	**X** L∃(p) : A Respectively **X** R∃(p) : φ(L(p))
X p : {x : A	φ}	**Y** ?$_L$ Or **Y** ?$_R$ **[the challenger has the choice]**	**X** L^{...}(p) : A Respectively **X** R^{...}(p) : φ(L(p))
	Y ?$_{prop}$	**X** (∀x : A)φ : prop
X p : (∀x : A)φ	**Y** L∀(p) : A	**X** R∀(p) : φ(L(p))
	Y ?$_{prop}$	**X** B(k) : prop
X p : B(k) (for atomic B)	**Y** ?	**X** sic (n) (X indicates that Y posited it at move n)

Note that in every rule there is first a challenge in the form "Y ?$_{prop}$". With this rule Y is asking X to confirm that the expression he posited is a proposition in accord with the formation rules.

Expressions L(p) and R(p) are called *instructions*. They are used when the form of play–objects is not evident immediately. They can be read as "take the left part of *p*" and "take the right part of *p*".

The informal explanations of the particle rules are the following:

1. Disjunction: if X posits an expression in the form of disjunction φ ∨ ψ, the play consists of two plays, p_1 and p_2, where X plays for φ in p_1 and for ψ in p_2. It is up to X to switch from p_1 to p_2, or the other way around, as he pleases.

2. Conjunction: if X posits an expression in the form of conjunction φ ∧ ψ, the play consists of two plays, p_1 and p_2, where X plays for φ in p_1 and for ψ in p_2. It is Y who can switch from p_1 to p_2 or the other way around.

3. Implication: if X posits an expression in the form of implication φ → ψ, then the play consists of two plays p_1 and p_2, where in p_1 it is Y who plays for φ and in p_2 it is X who plays for ψ.

4. Negation: if X posits an expression in the form of negation ¬φ the play consists of two plays, p_1 and p_2, where in p_1 it is Y who plays for φ and in p_2 it is X who plays for ⊥. It is up to X to switch from p_1 to p_2, or the other way around, as he pleases. The negation is understood in an intuitionistic manner, as the implication φ → ⊥. The negation rule amounts to a switch of roles for the players in a game.

5. The rules for quantifiers are somewhat more complicated. First, it should be noted that the quantifiers are written in the form (∃x : A)φ and (∀x : A)φ, as in CTT. This form expresses the CTT idea that a quantification is always relative to a set of a certain kind. Of course, we have to specify the kind of objects that form the set, and this is done in dialogues by means of the structural rule SR4.1., which will be explained later. This structural rule allows for "resolving the instructions," after which the kind of objects in question is specified. Second, the rules for quantifiers underline two important ingredients of the meaning of quantifiers: the choice of a value for a bound variable and the instantiation of a formula when the bound variable is replaced with the chosen value.

6. There is one last rule left to explain, namely that introduced between the rules for two quantifiers. It concerns a posit that X can make in the form {x : A | φ} which should be read as " those elements of A such that φ". Such expressions are used in CTT to deal with separate subsets. The idea behind such a posit is that there is at least one element $L^{(...)}(p)$ of A that is a witness for $φ(L^{(...)}(p))$. This expression cannot be challenged by the question "Y $?_{prop}$" because there is no assumption that the posit is a proposition: the expression stands here for a set.

There is a remark to be made here. Play objects for a conjunction and play objects for an existentially quantified expression are of the same form. The same stands for play objects for an implication and a universally quantified expression. There is a similarity between those logical constants in CTT where the conjunction and existential quantifier are two particular cases of the Σ operator, while the material implication and universal quantifier are two particular cases of the Π operator.[75]

There are some more rules that need to be introduced. First, *definitional equality*.

∀-Equality rule : if player **X** posits the function f(x) : B(x) (x : A) and if in a further move this function has been applied to the play-object k, then the challenger Y can force X to posit the equality $f(k) = R^f(k)$.

∃-Equality rule : If player **X** posits the pair of expressions k_1 : A and k_2 : $B(k_1)$ (as a response to a challenge on ∃-move) , the challenger Y can ask X to posit:

$$L^∃ = k_1 : A \text{ and/or } R^∃ = k_2 : B (k_1).[76]$$

The rules concerning the properties of sets are the following:

---

[75] For the full explanation of Σ and Π operators in CTT see Ranta (1994: chapter 2). For a discussion about the parallel with dialogical rules see Rahman, Clerbout (2014).
[76] See Rahman, Clerbout, Jovanovic, forthcoming.

Reflexivity within a *set*

Posit	Challenge	Defence
X-- A : set	**Y**-?$_{set}$- refl	**X**-- A = A : set

Symmetry within a **set**

Posit	Challenge	Defence
X-- A = B : set	**Y**-?$_B$- symm	**X**-- B = A : set

Transitivity within a **set**

Posit	Challenge	Defence
X-- A = B : set **X**-- B = C : set	**Y**-?$_A$- trans	**X**-- A = C : set

Reflexivity within **A**

Posit	Challenge	Defence
X-- a : A	**Y**-? a refl	**X**-- a = a: A

Symmetry within **A**

Posit	Challenge	Defence
X-- a = b : A	**Y**-?$_b$- symm	**X**-- b = a : A

Transitivity within **A**

Posit	Challenge	Defence
X-- a = b : A **X**-- b = c: A	**Y**-?$_a$- trans	**X**-- a = c : A

Set-equality (or extensionality)

Posit	Challenge	Defence
X-- A = B : set	**Y**-?$_{ext}$- a : A	**X**-- a : B

Posit	Challenge	Defence
X-- A = B : set	**Y**-?$_{ext}$- a = b : A	**X**-- a = b : B

Set-substitution

Posit	Challenge	Defence
X-- B(x) : set (x : A)	**Y**-?$_{subst}$ a = c : A	**X**-- B(a) = B(c) : set

Posit	Challenge	Defence
X-- C(x, y) : set (x : A, y : B(x))	**Y**-?$_{subst}$ a = c : A	**X**-- C(a, y) = C(c, y) : set

Posit	Challenge	Defence
X-- C(x, y) : set (x : A, y : B(x))	**Y**-?$_{subst}$ b = d: B(x)	**X**-- C(x, b) = C(x, d) : set

Posit	Challenge	Defence
X-- b(x) : B(x) (x : A)	**Y**-?$_{subst}$ - a = c : A	**X**-- b(a) = b(c) : B(a)

Before offering a presentation of structural rules there is one more rule that will be important for the treatment of branching quantifiers: the rule for the *function substitution*. This rule provides a means to distinguish between a provisional clause where a defender posits the expression within the clause and a provisional clause where a challenger does this. If it is the defender who is committed to the proviso we use the notation <...>; if it is the challenger, we use the notation (...). It is a novelty comparing to standard CTT because the rules are formulated in terms of game theory and there are the players involved. Here is the rule:

Posit	Challenge	Defence
$X \,!\, p : \varphi \, [f(k_1)]$	$Y \, f(k_1)/?$	$X \; p : \varphi \, [k_2 \,/\, f(k_1)]$ $< \varphi[f(k_1)] = \varphi \, [k_2 /\, f(k_1)]$ $: set]>$

Note that the defence requires the defender to posit both the substitution and the corresponding set equality. In other words, the defensive move is understood in the following manner: *f(k₁) is substituted by k₂, provided that X posits φ [f(k₁)] = φ [k₂/ f(k₁)] : set.* Thus, the rule could also be formulated as involving two challenges (and two defences). However, this might trigger two different plays (namely, when the challenger has chosen rank 1) and this might make the core quite complex (including variations in the order of the two challenges). The twofold formulation is the following:

Posit	Challenge	Defence
$X \,!\, p : \varphi \, [f(k_1)]$	$Y \, L^{\,f(k1)/}?$	$X \; p : \varphi \, [k_2 \,/\, f(k_1)]$
	$Y \, R^{\,f(k1)/}?$	$X \; <\varphi[f(k_1)] = \varphi[k_2/\, f(k_1)] : set]>$

II. 3. 4. Structural rules

As priviously mentioned, dialogical logic has a very subtle approach to meaning. The particle rules presented earlier provide the *local meaning*. Now I will present the *structural rules* where *global semantics* is achieved. These rules determine the general course of a dialogical game.

SR0 (Starting rule): A dialogical game begins with the thesis posited by the Proponent. Then both players choose their *repetition ranks*. A repetition rank determines the number of challenges a player can play related to the same move.

SR1i (Intuitionistic Development rule): After the thesis is posited and the repetition ranks chosen, the players continue playing one after the other. Each move is either a challenge or a defence in reaction to the previous move, and it is guided by the particle rules. In intuitionistic dialogues the players can answer only *the last non-answered* challenge of the adversary.

[**SR1c (Classical Development rule):** After the thesis is posited and the repetition ranks chosen, the players continue playing one after the other. Each move is either a challenge or a defence in reaction to the previous move and it is guided by the particle rules. In the classical dialogues the players can answer any challenge—even the one that has already been answered.]

SR2 (Formation first): The first challenge the Opponent launches is the request '?$_{prop}$'. Then the formation rules are applied and the Proponent must defend that the thesis is indeed a proposition. After this challenge, the Opponent can play following other particle and structural rules.

SR3 (Modified Formal rule): The Proponent is not allowed to challenge the Opponent's elementary sentences. The Opponent is allowed to challenge the Proponent's elementary sentences if the Opponent did not concede them himself earlier.

SR4.1 (Resolution of instructions): If a player posits a move with the instructions $I_1, ..., I_n$, his adversary can ask him to replace these instructions (or some of them) with suitable play-objects. Those play-objects are chosen by the defender of the instructions, except in the following cases: if there is a challenge of a universally quantified expression or implication and the instruction (or the list of instructions) occurs at the right of the colon and the posit is the tail of that universally quantified expression or the tail of the implication. In these cases it is the challenger who can choose play-objects. The rule is illustrated below:

Posit	Challenge	Defence
X $\pi(I_1, ..., I_n)$	Y $I_1, ..., I_m/?$ (m≤ n)	**X $\pi(b_1, ..., b_m)$** - if the instruction that occurs at the right of the colon is the tail of either a universal or an implication (such that $I_1,..., I_n$ also occur at the left of the colon in the posit of the head), then **$b_1,...,b_m$ are chosen by the challenger** - Otherwise **the defender chooses**

It can happen that instructions are embedded in the form: $I_1(...(I_k)...)$. In that case the rule implies that one must make the substitutions in the opposite direction: from I_k to I_1.

SR4.2 (Substitution of instructions): If one of the players chooses the play-object b for the instruction I, and player **X** posits $\pi(I)$ at some point, then Y can ask X to substitute I with b in any posit **X** $\pi(I)$. The rule is illustrated below:

Posit	Challenge	Defence
Player 1 $\pi_i(I)$ **Player 2** $I /?$ **Player 3** $\pi_i(b)$ ...**X**-- $\pi_j(I)$	**Y** ? b/I	**X** $\pi_j(b)$

Rules **SR4.1** and **SR4.2,** in which the play-objects are being substituted for the instructions, serve as a tool for interpreting, in a satisfactory manner, some puzzling natural-language sentences, such as those with anaphoric expressions. I mentioned in the first section of this book that Hintikka conceived IF logic as the true logic of natural language (that is, of the natural-language fragment that is usually considered first-order). One of his arguments was that at least some natural-language sentences need the explicit treatment of dependencies between quantifiers and the specification of their scope in order to be well interpreted. According to Hintikka and his associates, GTS is the most satisfactory semantics of anaphoric expressions. I will argue later that the explicit CTT language in the dialogical framework provides all the means needed to create an account of quantifier dependencies and scopes and to give a satisfactory interpretation of anaphora. The last chapter of this book will be dedicated to this problem.

SR5 (Winning rule for dialogues) If one of the players posits "p : ⊥" in a dialogue, he loses the play. Otherwise the player who makes the last move in the dialogue wins.

At the end it remains to precise that the formula is valid if there is a winning strategy for the Proponent in a dialogical game played for that formula. In material dialogues, the sentence at stake is satisfied in a given model.

The strategy is defined in a similar manner as in GTS: it is a function assigning a move to a player every time it is that player's turn to move in accordance with particle and structural rules.

Now I have enough of tools at my disposal. I will proceed to examine some of Hintikka's favorite accomplishments of GTS: the justification of the axiom of choice and the interpretation of anaphoric expressions.

PART III

Hintikka's take on the axiom of choice and the constructivist challenge

Now it is time to compare and contrast Martin-Löf's analysis of the axiom of choice with Jaakko Hintikka's standing on this axiom. Hintikka claims that GTS justifies Zermelo's axiom of choice in a first-order way that is perfectly acceptable for constructivists. I announced in the introduction to this book that I will use Martin-Löf's results to refute Hintikka's claims. Let me repeat that Martin-Löf's results lead to several important conclusions:

1) There is a reading of Hintikka's preferred version of the axiom of choice that makes it acceptable for constructivists, and under that reading the meaning of the axiom does not involve higher-order logic.

2) The reading that makes the axiom of choice acceptable for constructivists is based on an intensional take on functions. However, in the classical understanding of Zermelo's axiom of choice, the property of extensionality is assumed and this is the real reason behind the constructivists' rejection of it.

3) One further point that can be made is that rendering explicit dependence and independence patterns, which is Hintikka's main interest in constructing IF logic, is also possible within the CTT framework, but without paying the price of a system that is neither axiomatizable nor has an underlying theory of inference—let us not forget that inference is the heart of logic, after all. The possibility of accounting for these dependence and independence patterns comes from the CTT approach to meaning where (standard) meta-logical features are explicitly presented at the object-language level. In the preceding chapter of this book it was shown that recent developments in dialogical logic make clear that this approach to meaning in general and to the axiom of choice in particular is very natural to game-theoretical approaches. In what follows, I want to vindicate Hintikka's plea for the fruitfulness of game-theoretical semantics in the context of the foundation of mathematics. However, this has to be done in a manner quite different to that which he proposes. In fact, from the dialogical point of view, those actions that constitute the

meaning of logical constants, such as choices, are the crucial element of its full-fledged local semantics.

III. 1. Recapitulation: Hintikka's take on the axiom of choice

As explained in the first chapter of this book, in GTS truth and falsity conditions are defined via notion of a winning strategy. A strategy is defined by a finite set of choice functions or *Skolem functions*. Values of functions indicate which individuals the Verifier has to choose in his actions when the game is played for an existential quantifier (or which disjunct, when the game is played for a disjunction) in order to win the game, and which individuals the Falsifier has to choose when the game is played for a universal quantifier (or which conjunct, when the game is played for a conjunction). Remember the example from section *I.5.2.*:

1) $\forall x \exists y \; C(x,y)$

This sentence is true if there is a winning strategy for Eloise. The winning strategy shows her how to select a value for y in function of the value of x ($f(x) = y$). The existence of such a strategy is confirmed in the following manner:

2) $\exists f \forall x \; C(x,f(x))$

If 1) and 2) are linked with a conditional, the following formulation of the *axiom of choice* results:

3) $\forall x \exists y \; C(x,y) \rightarrow \exists f \forall x \; C(x,f(x))$

From the GTS point of view, the truth of 3), expressing a version of the axiom of choice, is derived from the very definition of truth. In fact, as I will discuss further on, it is related to the truth of the universal.

Let me repeat that the axiom of choice is essential for GTS, since without it Tarski-style semantic and GTS for classical first-order logic would not be equivalent. The reason for this is that the strategies in GTS are understood as *deterministic* strategies which impose choices on the Verifier and Falsifier, leaving them with no real options. However, Hintikka insists that there is nothing troubling about the axiom of choice and that it actually constitutes the very conception of truth. Recall the quotation given above:

This paradigm problem concerns the status of the axiom of choice. This axiom was firmly rejected by Brouwer and it was mooted in the controversies between the French intuitionists and their opponents....The axiom of choice is true. The idea of "choosing" or "finding" suitable individuals is systematised in what is known as game-theoretical semantics. For mathematicians, this semantics is no novelty, however, but little more than a regimentation and generalisation of the way of thinking that underlies mathematicians' classical (or perhaps I should say Weierstrassian) epsilon-delta analyses of the basic concepts of calculus, such as continuity and differentiation...

...To return to the usual axiom of choice, it is thus seen to be unproblematically true. How can any intuitionist deny the axiom of choice...? What can possibly go wrong here? Moreover, evoking the concept of knowledge, either in the form of epistemic logic or informally, does not seem to help an intuitionistic critic of the axiom of choice at all, either. The discussion of the axiom of choice between intuitionists and classicists has conducted at cross-purposes. It can only be dissolved by making distinction between knowing that and knowing what that neither party has made explicit. (Hintikka, 2001)

Zermelo did not begin to axiomatize set theory unselfishly from the goodness of his theoretical heart. His main purpose was to justify his well-ordering theorem. In practice, this largely meant to justify the axiom of choice. [...]. But that is not the full story. Worse still: Zermelo's specific enterprise was unnecessary, in that the so- called axiom of choice turns out to be in the bottom a plain first-order logical principle. (Hintikka, 2011)

In fact, as I will show below, the last sentence is true: the truth of the axiom of choice involves only first-order logic. However, in contrast to what Hintikka argues against the intuitionists, the axiom of choice holds in the context of intuitionistic first-order logic enriched with the axioms of ZF-set theory, formulated in the style of CTT.

III.2. Martin-Löf on the axiom of choice

Let me give a brief outline of the axiom of choice. It is well known that it was first introduced by Zermelo in 1904 in order to prove the well-ordering theorem.[77] Zermelo gave two formulations of this axiom, the first in 1904 and the second in 1908. The second formulation is relevant for this discussion, since it is related to both Hintikka and Martin-Löfs' formalizations:

> *A set S that can be decomposed into a set of disjoint parts A, B, C, ... each of the containing at least one element, possesses at least one subset S_1 having exactly one element with each of the parts A, B, C, ... considered.* (Zermelo, 1908)

The Axiom immediately attracted attention and both of its formulations were criticized by constructivists such as Baire, Borel, Lebesgue, and Brower. The first objections were related to the non-predicative character of the axiom, because a choice function was supposed to exist without it being shown constructively that it did. However, the axiom found its way into ZFC set theory and was finally accepted by the majority of mathematicians because of its usefulness in different branches of mathematics. Indeed, it is probably the most fruitful axiom of set theory, with at least two hundred logical consequences that are formally equivalent to it.[78]

In 1938 Gödel proved the relative consistency of the axiom of choice with ZF set theory by constructing the inner model of constructible sets where the axiom of choice is satisfied.[79] In 1963 Cohen used his famous method of *forcing* to prove the independence of the axiom of choice from the rest of ZF set theory.[80]

Martin-Löf produced a proof of the axiom in a constructivist setting by bringing together two seemingly incompatible perspectives on this axiom, namely:

> 1) Bishop's surprising observation from 1967, namely that *a choice function exists in constructive mathematics, because a choice is implied by the very meaning of existence.*

---

[77] The theorem states that every set can be well–ordered.
[78] See Bell (2009: 3).
[79] Gödel (1939).
[80] Cohen (1963).

2) The proof by Diaconescu from 1975 and by Goodman and Myhle from 1978 that the Axiom of Choice implies the law of the excluded middle.

Martin-Löf (2006) shows that there are indeed some versions of the axiom of choice that are perfectly acceptable for constructivists, namely those where the choice function is defined *intensionally*. In order to see this, the axiom must be formulated within the frame of a CTT-setting. Indeed, such a setting allows comparing the extensional and the intensional formulation of the axiom. It is in fact the extensional version that implies the law of the excluded middle, whereas the intensional version is compatible with Bishop's remark. As Martin-Löf (2006: 349) puts it:

> [...] *this is not visible within an extensional framework, like Zermelo-Fraenkel set theory, where all functions are by definition extensional."*.

In CTT the truth of the axiom actually follows rather naturally from the meaning of quantifiers:

Take the proposition $(\forall x: A)\ B(x)$ where $B(x)$ is of the type proposition provided that x is an element of the set A. If the proposition is true; then there is a proof for it. Such a proof is in fact a function that for every element x of A renders a proof of $B(x)$. This is how Bishop's remark should be understood: the truth of a universal amounts to the existence of a proof, and this proof is a function. Thus, the truth of a universal amounts, in the constructivist account, to the existence of a function. The proof of the axiom of choice can be developed from this quite straightforwardly. If we recall that in the CTT-setting:

1) the existence of a function from A to B amounts to the existence of the proof-objects for the universal *every A is B*, and

2) the proof of the proposition Bx, existentially quantified over the set A, amounts to a pair such that the first element of the pair is an element of A and the second element of the pair is a proof of Bx.

The full-fledged formulation of the axiom of choice—where we make explicit the set over which the existential quantifiers are defined—is the following:

$$(\forall x : A)\,(\exists y : Bx)\ C(x,y) \rightarrow (\exists\ f : (\forall x : A)\ Bx)\,(\forall x : A)\ C(x, f(x))$$

The proof of Martin-Löf (1980, pp. 50–51) is the following:

> *The usual argument in intuitionistic mathematics, based on the intuitionistic interpretation of the logical constants, is roughly as*

fallows: to prove $(\forall x)(\exists y)C(x,y) \rightarrow (\exists f)(\forall x)C(x,f(x))$, assume that we have a proof of the antecedent. This means we have a method which, applied to an arbitrary x, yields a proof of $(\exists y)C(x,y)$. Let f be the method which, to an arbitrarily given x, assigns the first component of this pair. Then $C(x,f(x))$ holds for an arbitrary x, and hence, so does the consequent. The same idea can be put into symbols getting a formal proof in intuitionistic type theory. Let A : set, B(x): set (x: A), C(x,y): set (x: A, y: B(x)), and assume z: ($\Pi x: A$)($\Sigma y: B(x))C(x,y)$. If x is an arbitrary element of A, i.e. x: A, then by Π- elimination we obtain

Ap(z,x): ($\Sigma y: B(x))C(x,y)$

We now apply left projection to obtain

p(Ap(z,x)): B(x)

and right projection to obtain

q(Ap(z,x)): C(x,p(Ap(z,x))).

By λ-abstraction on x (or Π- introduction), discharging x: A, we have

(λx) p(Ap(z,x)): ($\Pi x: A)B(x)$

and by Π- equality

Ap((λx) p(Ap(z,x), x) $=$ p(Ap(z,x)): Bx.

By substitution [making use of C(x,y): set (x: A, y: B(x)),] we get

C(x, Ap((λx) p (Ap(z,x), x) $=$ C(x, p(Ap(z,x)))

[that is, C(x, Ap((λx) p (Ap(z,x), x) $=$ C(x, p(Ap(z,x))): set]

and hence by equality of sets

q(Ap(z,x)): C(x, Ap((λx) p (Ap(z,x), x)

where ((λx) p (Ap(z,x)) is independent of x. By abstraction on x

((λx) p (Ap(z,x)): ($\Pi x: A)C(x, Ap(($\lambda x$) p (Ap(z,x), x)

We now use the rule of pairing (that is Σ- introduction) to get

(λx) p(Ap(z,x)), (λx) q(Ap(z,x)): ($\Sigma f: (\Pi x: A)B(x))(\Pi x: A)C(x, Ap(f,x))$

(note that in the last step, the new variable f is introduced and substituted for

((λx) p (Ap(z,x)) in the right member). Finally by abstraction on z, we obtain

(λz)((λx) p (Ap(z,x)), ((λx) q (Ap(z,x)): (Πx: A)(Σy: B(x))C(x,y)→

(Σf: (Πx: A)B(x))(Πx: A)C(x, Ap(f,x)).

Curiously, this seems to be close to Hintikka's own formulation and even to his analysis that a winning strategy for a universal amounts to the existence of a (Skolem) function. This is curious, since Martin-Löf's proof is developed within a constructivist setting. Moreover, what is shown by Martin-Löf (2006) is that what is wrong with the axiom, from the constructivist point of view, is its extensional formulation. That is:

$$(\forall x : A) (\exists y : Bx) C(x,y) \rightarrow (\exists f : (\forall x : A) Bx) (Ext(f) \& (\forall x : A) C(x, f(x))$$

Where $(Ext(f) = ((\forall i, j : A) (i =_A j \rightarrow f(i) = f(j))$

Therefore, from the constructivist point of view, what is really wrong with the classical formulation of the axiom of choice is the assumption that from the truth that *all of the A are B* we can obtain a function that satisfies the property of extensionality. In fact, as shown by Martin-Löf (2006), the classical version holds, even constructively, if we assume that there is only one such choice function in the set at stake:

$$(\forall x : A) (\exists! y : Bx) C(x,y) \rightarrow (\exists f : (\forall x : A) Bx) (Ext(f) \& (\forall x : A) C(x, f(x))$$

Let us retain that:

1) If we take $(\forall x : A) (\exists y : Bx) C(x,y) \rightarrow (\exists f : (\forall x : A) Bx) (\forall x : A) C(x, f(x))$ to be the formalization of the axiom of choice, then that axiom is not only unproblematic for a constructivist but is also a theorem. But this formalization is a full-fledged formulation of the version Hintikka adopts.[81] Certainly, the point is that the CTT-formulation stresses

---

[81] Indeed, Martin-Löf's formalization follows from making explicit in Hintikka's formulation $\forall x \exists y \ C(x,y) \rightarrow \exists f \forall x \ C(x,f(x))$ the range of its quantifiers, that is: $\forall x$ quantifies over, say the set A, $\exists y$ quantifies over, say the set Bx, and $\exists f$, over the set $(\forall x : A) Bx$.

explicitly that the choice function at stake has been defined by means of intensional equality, but Hintikka seems to assume extensionality. In fact it is the CTT-explicit language that allows a fine-grained distinction between, on the surface, the equivalent formulations. This is due to the expressive power of CTT that allows expressing at the object-language level properties that in other settings are left implicit in the metalanguage. This leads us to the second point;

2) According to the constructivist approach, functions are identified with proof-objects for propositions and are given in the *object-language,* as objects of a certain type. Understood in that way, functions belong to the lowest level of entities and there is no jumping to a higher order. Once more, the truth of a first-order universal sentence amounts to the existence of a function that is defined by means of the elements of the set over which the universal quantifies and by the first-order expression Bx. The existence of such a function is the CTT way of expressing at the object language level that a given universal sentence is true.

Thus, Hintikka is right in defending the idea that we only need a first-order language, but this does not really support his attachment to the classical understanding of the axiom of choice.

III.3. The axiom of choice in the dialogical framework

Let us use now the framework we have presented above, where the dialogical approach to logic is linked with CTT. Shahid Rahman and Nicolas Clerbout (2014) worked out a dialogical proof of the constructive formulation of the axiom of choice. They thus proved Martin-Löf's point in a game-theoretical framework. The proof is as follows:[82]

First play: Opponent's 9th move asks for the left play object for the existential quantification on f.

---

[82] See Rahman, Clerbout (2014). It will also appear in Rahman, Clerbout, Jovanović, forthcoming. In the present proof two plays are extracted from the extensive tree of all the plays-those which constitute the so-called core of the strategy (that is, of the dialogical proof). The proof that these plays render the corresponding CTT demonstration can be found in Rahman, Clerbout (2014).

	O			P	
	H1: $C(x, y)$: set $(x : A, y : B(x))$ H2: $B(x)$: set $(x : A)$			$p : (\forall x : A) (\exists y : B(x)) C(x,y) \rightarrow (\exists f : (\forall x : A) B(x)) (\forall x : A) C(x, f(x))$	0
1	m:= 1			n:= 2	2
3	$L^{\rightarrow}(p)$: $(\forall x : A) (\exists y : B(x)) C(x,y)$	0		$R^{\rightarrow}(p)$: $(\exists f : (\forall x : A) B(x)) (\forall x : A) C(x, f(x))$	6
5	v : $(\forall x : A) (\exists y : B(x)) C(x,y)$		3	$L^{\rightarrow}(p)$ /?	4
7	$R^{\rightarrow}(p)$ /?	6		(v, r) : $(\exists f : (\forall x : A) B(x)) (\forall x : A) C(x, f(x))$	8
9	$L_?$	8		$L^{\exists}(v, r)$: $(\forall x : A) B(x)$	10
11	$L^{\exists}(v, r)$ /?	10		v : $(\forall x : A) B(x)$	12
13	$L^{\forall}(v)$: A	12		$R^{\forall}(v)$: $B(w)$	26
15	$w : A$		13	$L^{\forall}(v)$: /?	14
19	$R^{\forall}(v)$: $(\exists y : B(w)) C(w, y)$		5	$L^{\forall}(v)$: A	16
17	$L^{\forall}(v)$ /?	16		$w : A$	18
21	(t_1, t_2) : $(\exists y : B(w)) C(w, y)$		19	$R^{\forall}(v)$ /?	20
23	$L^{\exists}((t_1, t_2)$: $B(w)$		21	$L_?$	22
25	$t_1 : B(w)$		23	$L^{\exists}(t_1, t_2)$ /?	24
27	$R^{\forall}(v)$ /?	26		$t_1 : B(w)$	28

Description:

Move 3: *After setting the thesis and establishing the repetition ranks* **O** *launches an attack on material implication.*

Move 4: P *launches a counterattack and asks for the play object that corresponds to* $L^{\to}(p)$.

Moves 5, 6: O *responds to the challenge of 4. P posits the right component of the material implication.*

Moves 7, 8: O ' *asks for the play object that corresponds to* $R^{\to}(p)$. **P** *responds to the challenge by choosing the pair (v, r) where v is the play object chosen to substitute the variable f and r the play object for the right component of the existential.*

Move 9: O *has here the choice to ask for the left or the right component of the existential. The present play describes the development of the play triggered by the left choice.*

Moves 10-26: *follow from a straightforward application of the dialogical rules. Move 26 is an answer to move 13, since P decided to have enough information to apply the characteristic –copy-cat method imposed by the formal rule.*

Move 27-28: O *asks for the play object that corresponds to the instruction posited by P at move 26 and P answers and **wins** by applying copy-cat to O's move 25. Notice that 28 this is not a case of function substitution: it is simply the resolution of an instruction.*

Second play: *Opponent's 9th move asks for the right play object for the existential quantification on f.*

O			P	
H1: $C(x, y)$: set (x : A, y : B(x)) H2: $B(x)$: set (x : A)			$p : (\forall x : A) (\exists y : B(x)) C(x,y) \to (\exists f : (\forall x : A) B(x)) (\forall x : A) C(x, f(x))$	
1	m:= 1			n:= 2
3	$L^{\to}(p)$: $(\forall x : A) (\exists y : B(x)) C(x, y)$	0		$R^{\to}(p)$: $(\exists f : (\forall x : A) B(x)) (\forall x : A) C(x,f(x))$
5	v : $(\forall x : A) (\exists y : B(x))$		3	$L^{\to}(p)$ /?

	$C(x, y)$				
7	$R^{\rightarrow}(p)$ /?	6		$(v, r) : (\exists f : (\forall x : A) B(x)) (\forall x : A) C(x, f(x))$	8
9	$R_?$	8		$R^{\exists}(v, r) : (\forall x : A) C(x, L^{\exists}(v, r)(x))$	10
11	$L^{\exists}(v, r)$ /?	10		$R^{\exists}(v, r) : (\forall x : A) C(x, v(x))$	12
13	$R^{\exists}(v, r)$ /?	12		$r : (\forall x : A) C(x, v(x))$	14
15	$L^{\forall}(r) : A$	14		$R^{\forall}(r) : C(x, v(w))$	32
17	$w : A$		15	$L^{\forall}(r) : $ /?	16
21	$R^{\forall}(v) : (\exists y : B(w)) C(w,y)$		5	$L^{\forall}(v) : A$	18
19	$L^{\forall}(v)$ /?	18		$w : A$	20
23	$(t_1, t_2) : (\exists y : B(x)) C(x ,y)$		21	$R^{\forall}(v)$ /?	22
25	$L^{\exists}((t_1, t_2): B(w)$		23	$L_?$	24
27	$t_1 : B(w)$		25	$L^{\exists}(t_1, t_2)$ /?	26
29	$R^{\exists}(t_1, t_2): C(w, t_1)$		23	$R_?$	28
31	$t_2 : C(w, t_1)$		29	$R^{\exists}(t_1, t_2)$ /?	30
33	$R^{\forall}(r)$ /?	32		$t_2 : C(w, v(w))$	34
35	$v(w)$ /?	34		$t_2 : C(w, t_1)$ $< C(w, t_1) = C(w, t_1 / v(w)) : set >$	42
41	$C(w, t_1) = C(w, t_1 / v(w)) : set$		H1? *subst*	$v(w) = t_1 : B(w)$	36
37	$v(w) = t_1 : B(w)$?	36		sic (39)	40
39	$v(w) = t_1 : B(w)$		5,18,	? $\forall$-eq.	38

		21,25	

Description:

Move 9: *Until move 9 this play is the same as the previous. In the present play, in move 9 the Opponent chooses to ask for the right-hand side of the existential posited by **P** at 8.*

Moves 10-34: *the Proponent substitutes the variable f by the instruction correspondent to the left-hand component of the existential, i.e., $L^{\exists}(v, r)$. By this **P** accounts for the dependence of the right-hand part on the left-hand component. The point is that the local meaning of the existential requires this dependence of the right component to the left component even if in this play the Opponent, due to the restriction on rank 1, she can ask only for the right-hand part*

*The conceptually interesting moves start with 35, where the opponent asks to **P** to substitute the function. As already pointed out, in order to respond to 35 the opponent's move 31 is not enough. Indeed the proponent needs also to posit $C(w, t_1) = C(w, t_1 / v(w))$: set. **P** forces **O** to concede this equality (41), on the basis of the substitutions w / x and t_1 / y on H1 (we implemented the substitution directly in the answer of **O**) given the ∀-equality $v(w) = t_1$ in B(w) (36), and given that this ∀-equality yields the required set equality. Moreover, **P**'s posit of the ∀-equality (36) is established and defended by moves 38-40.*

Thus, in the dialogical framework the point is shown straightforwardly. The axiom of choice is valid in a constructive setting only if the choice function is defined intensionally.

A winning strategy for the Proponent in a dialogical game for the implication at stake follows immediately from the very meaning of the antecedent. The Proponent wins the play because he copies the choice that the Opponent made for y in the antecedent and uses it for f(x) in the consequence. This is possible because both y and f(x) are objects of the same type: B(x), given that x: A. It is important to notice that C(x, y) in the antecedent and C(x, f(x)) in the consequence are two *intensionally* equal sets for every x: A. It is thus the intensionality that is sine qua non of the proof of the axiom of choice in a constructive setting, unless the uniqueness of the function is assumed.

Let us entertain for the moment the idea that C(x,y) were defined *extensionally*. In that case, for any extensionally equal a and b in A, we can substitute a: A occurring at the place of x in C(x, y) with b: A; and the same goes for every c: Bx (x : A), as pointed out in Rahman, Clerbout (2014, appendix). Rahman and Clerbout (Ibid.) work out the dialogical rule for extensionality of function in the following way:

Posit	Challenge	Defence
X p: $Ext(f)$ (where $f: A \rightarrow B$)	**Y-** $L^{Ext(f)}(p: k_i =_A k_j$ (where k_i and k_j are elements of A chosen by the challenger)	**X** $R^{Ext(f)}(p: f(k_i) =_B f(k_j)$

The idea underlying the rule should be clear: The play object p is constituted by a pair, the left and the right side of p. If X posited that f (f : A$\rightarrow$B) is extensional, then the left part of p is a play object for the extensional equality (in A) - posited by Y - of two elements of A chosen by the challenger. The, right part of p is a play object for the extensional equality (in B) – posited by X - of the two functions that take as arguments the elements of A chosen by the challenger.

We will also need also a rule for substitution in extensional relations:

Posit	Challenge	Defence
.		
X- $p : Ext(C(k_i, k_z))$ $(k_i : A, k_z : B)$	**Y-** $L_1^{Ext(Cx,y)}(p) : k_j =_A k_i$	**X-** $R_1^{Ext(Cx,y)}(p)C(k_j, k_z)$
X- $p : Ext(C(k_i, k_z))$ $(k_i : A, k_z : B)$	**Y-** $L_2^{Ext(Cx,y)}(p) : k_w =_B k_z$	**Y-** $R_2^{Ext(Cx,y)}(p) : C(k_i, k_w)$

Now, the point is that unless the unicity of the function is assumed, in a dialogical game for the extensional formulation of the axiom of choice the Proponent cannot have a winning strategy because, even when Opponent concedes that there is a function at stake (as he does in the intensional formulation of the axiom), he cannot be forced to concede the extensionality of that function. The extensional equality of two functions $f(a)$ and $f(b)$, for some a and b chosen by the Opponent (extensionally equal in A), can be obtained from the earlier concessions of the Opponent in the game only under the assumption that there is only one such function, i.e., if the Opponent concedes that there is only one element of the set Bx. In the last case, when the unicity of the function is assumed, there is a dialogical proof for the validity of the extensional version of the axiom of choice worked out by Rahman and Clerbout (2014):

	O		**P**	
	H_1: t : $(\forall x : A)$ $(\exists y : Bx)$ $C(x,y) \rightarrow (\exists f : (\forall x : A)$ $Bx)$ $(\forall x : A)$ $C(x, f(x))$ H_2: $L^3(R^{\rightarrow}(t)) = L^3(R^{\rightarrow}(p)) = g$: $(\forall x : A)$ Bx H_3: $Ext(C)$ (where C occurrs in g)		p : $(\forall x : A)$ $(\exists ! y : Bx)$ $C(x,y)$ $\rightarrow$ $(\exists f : (\forall x : A)$ $Bx)$ $Ext(f)$ $\wedge$ $(\forall x : A)$ $C(x, f(x))$	0
1	$m := 1$		$n := 2$	2
3	$L^{\rightarrow}(p)$: $(\forall x : A)$ $(\exists ! y : Bx)$ $C(x, y)$	0	$R^{\rightarrow}(p)$: $(\exists f : (\forall x : A)$ $Bx)$ $Ext(f)$ $\wedge$ $(\forall x : A)$ $C(x,f(x))$	6
5	g : $(\forall x : A)$ $(\exists ! y : Bx)$ $C(x, y)$	3	$L^{\rightarrow}(p)$ /?	4
7	$R^{\rightarrow}(p)$ /?	6	(g, r) : $(\exists f : (\forall x : A)$ $Bx)$ $Ext(f)$ $\wedge$ $(\forall x : A)$ $C(x, f(x))$	8
9	$R_?$	8	$R^3(g, r)$: $Ext(L^3(g, r)$ $\wedge$ $(\forall x : A)$ $C(x, L^3(g, r)(x)$	10
11	$L^3(g, r)$ /?	10	$R^3(g, r)$: $Ext(g)$ $\wedge$ $(\forall x : A)$ $C(x, g(x))$	12
13	$R^3(g, r)$ /?	12	r : $Ext(g)$ $\wedge$ $(\forall x : A)$ $C(x, g(x))$	14
15	L?		$L^{\wedge}(r)$: $Ext(g)$	16

17	$L^\wedge(r)$ / ?		$s : Ext(g)$	18
19	$L^{Ext(g)}(s) : a =_A b <a : A, b : A>$	18	w/ $R^{Ext(g)}(s) : g(a) =_{Ba} g(b)$	36
21	$R^\to(t) : (\exists f : (\forall x : A) Bx) (\forall x : A) C(x,f(x))$	H_1	$L^\to(t) : (\forall x : A) (\exists !y : Bx) C(x, y)$	20
23	... u/ $R^\exists(g,u): (\forall x : A) C(x, g(x))$	H_2	... (copy-cat 7-14)	22
25	$v / R^\forall(u) : C(a, g(a))$	23	$a / L^\forall(u) : A$	24
27	$v / R^\forall(u) : C(b, g(b))$	23	$b / L^\forall(u): A$	26
29	$v : Ext(C(b, g(b)))$	27, H_3	? $Ext(C)$	28
31	$v : C(a, g(b))$	29	$v : a =_A b$	30
33	$g : (\forall x : A)(\forall y: Bx) (\forall z: Bx) C(x,y) \wedge C(x, z) \to y =_{Bx} z$	5	- ? $\exists$!	32
35	w/ $R^\to(g) : g(a) =_{Ba} g(b)$	33	$v / L^\to(g) : C(a, g(a)) \wedge C(a, g(b))$ $(a : A, g(a), g(b) : Ba)$	34

Description:

Move 3: *After setting the thesis and establishing the repetition ranks **O** launches an attack on material implication.*

Move 4: ***P** launches a counterattack and asks for the play object that corresponds to $L^\to(p)$.**Moves 5, 6: O** responds to the challenge of 4 . **P** posits the right side of the material implication.*

Moves 7, 8: O *asks for the play object that corresponds to $R^\to(p)$. **P** responds to the challenge by choosing the pair (v, r) where v is the play object choosen to substitute the variable f and r the play object for the right side of the existential.*

Move 9: O *has here the choice to ask for the left or the right side of the existential. The present play describes the development of the play triggered by the right choice.*

Moves 10-18: *follow from a straightforward application of the dialogical rules.*

Move 19*: O launches an attack to the extensionality of g, choosing the element a and b from A and positing its extensional equality in A.*

Moves 20-23*: P forces O to posit u/ $R^3(g,u)$: ($\forall x$: A) C(x, g(x)). In fact, as mentioned above, the moves are a copy-cat of the moves -7-14 that force P to posit r/ $R^3(g,r)$: ($\forall x$: A) C(x, g(x)). In other words, P forces O to concede by the intensional axiom of choice (H_1) that there is a function g such that ($\forall x$: A) C(x, g(x)).*

Moves 24-26*: P forces to concede both C(a, g(a)) and C(b, g(b)), choosing precisely a, b (the elements of A chosen by O before).*

Moves 28-31*: P forces O to also concede C(a, g(b)) by making use of the extensionality of C (H_3).*

Moves 32-35*: because of moves 25 and 27 P has all what he needs to apply unicity and force O to concede that g(a) and g(b) are extensionaly the same function in Ba. We did not display here the possible (unsuccessful) counterattack of O on the conjuction C(a, g(a)) $\wedge$ C(A, g(b))since this posit is composed by copy-cat moves of P from moves 25 and 27 of O.*
Move 36*: the just conceded extensional equality g(a) $=_{Ba}$ g(b), can be now posited by P as an answer to the challenge of O launched at move 19, and win!*

At this point, I want to mention one recent publication of Jonathan Sterling (2015) concerning the axiom of choice, which is relevant for my discussion[83]. Sterling explores Diaconescu's proof that the axiom of choice in an extensional framework yields the law of the excluded middle[84]. As previously mentioned, when Martin- Löf was confronted with tension between Diaconescu's

---

[83] Thanks to Mark Van Atten who pointed out to me this very recent paper and thanks to Professor Rahman for his clarifications.
[84] Which Sterling labels *the taboo.*

theorem and Bishop's remark about the evident logical validity of the axiom of choice in constructive mathematics, he came up with a distinction between intensional and extensional version of the axiom, arguing that the former is the genuine theorem of CTT, while the later implies the law of the excluded middle. Martin- Löf's conclusion was that *extensionality does not come for free.* I followed in this chapter his intensional, intuitionistic perspective to develop my critique of Hintikka's take on the axiom of choice.

Sterling took another road in his paper, working in extensional Constructive Type Theory framework to argue that: 1) the culprit for the failure of Zermelo's axiom of choice is not the extensionality but rather an inept interpretation of quantifiers and 2) that the extensional CTT is still not susceptible to Diaconescu's theorem. More precisely, Sterling clames that the extensional version of the axiom of choice, as defined by Martin- Löf, operates on *setoids*[85], while his intensional version of the axiom operates on sets. What is wrong with the extensional version from this perspective is not the extensionality but an unjustified generalization that happens when one wants to pass from a function in the antecedent, defined on particular setoids, to a function that ranges over arbitrary quotient sets in the consequence. As Sterling puts it:

> *In this way, the confusion of ideas that leads one to believe in AC_{ext} is precisely (not just figuratively) the same confusion of ideas which leads one to believe in PEM[86] (if it can be said that anyone really believes in that anyway).*

This interesting point developed by Sterling does not really affect any of proofs previously given. Still we can entertain the idea that when working in the extensional CTT, the culprit for the failure of Zermelo's axiom is not extensionality but the unfounded generalization and we can examine how this reflects on Hintikka's standing. To put it straightforward, even this result does not come easy for Hintikka. As he defended the realist perspective where the existence of the choice function must be independent of players' knowledge, it might be arguable that with this assumption the unwanted generalization actually happens: the function in the consequence ranges over arbitrary setoids or arbitrary quotient set, independently of how it was determined in the antecedent. The conclusion is that in both intensional and extensional reading of Hintikka's formulation of the axiom of choice it is not possible to defend classical understanding of it, whether it is intensionality or extensionality that *does not come for free.*

---

[85] *Setoid* is a set together with an equivalence relation.
[86] The principle of the excluded middle.

III 4. Conclusion: The antirealist rejoinder of Hintikka's take on the axiom of choice

According to Hintikka, his preferred formulation of the axiom of choice, namely: $\forall x \exists y \, C(x,y) \rightarrow \exists f \forall x \, C(x,f(x))$,[87] is perfectly acceptable to constructivists. Let us recall that the GTS reading of its truth amounts to the existence of a winning strategy for Eloise in the game $G(\forall x \exists y \, C(x,y))$. The latter amounts to finding a "witness individual" y, dependent upon x, such that $C(x,y)$ is true—notice how close this formulation is to Martin-Löfs' "informal" description of the proof of the axiom quoted in the preceding section. In other words, the existence of a winning strategy for the game provides *a proof that the proposition S(x,y) is true in the model.*[88] Hintikka (1996a: 212) argues that it is the GTS reading that makes the axiom of choice acceptable to constructivists:

> *Moreover, the rules of semantical games should likewise be acceptable to a constructivist. In order to verify an existential sentence $\exists x S[x]$ I have to find an individual b such that I can (win in the game played with) S[b]. What could be a more constructivistic requirement than that? Likewise, in the verification game G(S1 v S2) connected with a disjunction (S1 v S2), the verifier must choose S1 or S_2 such that the game connected with it (i.e., G(S1) or G(S2)) can be won by the verifier. Again, there does not seem to be anything here to alienate a constructivist.*

As mentioned above, Hintikka is right in thinking that this formulation is acceptable to constructivists, but the reason for this is the underlying intensionality of the choice function. Hintikka would like to have a logic that can render classical mathematics. However, this is only possible if we assume extensionality. To have both extensionality and the constructive axiom of choice—without assuming the unicity of the choice function—seems impossible.

Hintikka's intention was to offer a realist foundation of mathematics on the first-order level in such a way that all classical mathematics can be formulated and it is still acceptable for constructivists. The preceding analysis shows that this amounts to having one's cake and eating it too. As pointed out by Sundhom (2013), under the constructivist reading, IF logic is granted the status of first-order logic,

---

[87] It is left implicit that $\forall x$ quantifies over some set A, $\exists y$ quantifies over some set B, and $\exists f$, over the set $((\forall x : A) \, Bx)$.

[88] See Hintikka (1996a, ch 2).

but in that case not all of classical mathematics can be saved. Moreover, if we comply with the constructive reading of IF logic, assuming that such a reading is possible, this makes the whole endeavour somewhat superfluous. Then we can use a version of constructive logic without getting into trouble with a non-axiomatisable logic.

Still, things might be even more complicated. The question is whether a constructivist reading of GTS is available to Hintikka at all.[89] That is, what is required for constructive reading is that functions are defined via intensional equality, but for that suitable formal equality rules are requisite. Such a formal rule is essential in a dialogical proof of the axiom—the one allowing the Proponent to rerun in the consequence the choice the Opponent made in the antecedent. This formal step is the guarantee that the function that makes the antencedent true is exactly the same function found in the consequence. The problem is that Hintikka refuses to have a formal rule, so it seems that the intensional version is simply not obtainable in his GTS. Another issue with the constructive reading of GTS is one I have mentionned before: the assymetry in the treatment of atomic and complex formulas.

Let's remind ourselves of a point made earlier in section *I.7*. Hintikka argues that the axiom of choice should already be acceptable for a constructivist under the GTS reading. However, if one wants to take this further, one can place a restriction on players' strategies to keep them to *recursive ones* (even if Hintikka himself is not in favor of such a restriction, at least not when it comes to descriptive role of logic in mathematics). The above analysis confirms what was said in *I.7.*: to obtain a constructivist reading of GTS in general and of the axiom of choice in particular, it is not the recursivity of function that is required. A game-theoretical approach acceptable to constructivists must have:

1. The constructive rule for negation, where $\neg A$ is understood as $A \rightarrow \bot$;

2. The means to express the force sentences have in language-use (assertions, questions, etc.);

3. The intensional reading of the choice function in the formulation of the axiom of choice. The previous arguments show that for this reading some kind of formal rule defining the intensional equality of choice function is required.

This argumentation leaves Hintikka in an unpleasant situation. It shows that IF logic is indeed a first-order logic only under a constructive reading which

[89] I thank for this remark to Prof. Dr Shahid Rahman, who made it during the Viva of my Thesis.

guarantees that functions are at the lowest level in the type hierarchy, but it is questionable whether this reading is available to Hintikka at all in the lack of a formal rule defining the equality of functions in GTS and given the assymetry in the treatment of atomic and complex formulas. Even if such a reading is somehow possible, we are left with the question of why we should use IF logic that lacks the theory of inference instead of some version of a constructive logic. In any case, the argumentation shows that Hintikka has to give up his dream of a *realist–nominalist* foundation of mathematics with his IF logic—if he holds on to realism, he has to accept jumping to the higher-order; otherwise, he has to comply with some kind of anti-realism.

Finally, let us recall once more one of the fundamental features of the dialogical approach to CTT: that all the actions that constitute the meaning are rendered explicit in the object-language. The game-theoretical reading of the axiom of choice stresses one of the more salient characteristics of CTT language: the judgement that a proposition is true can be expressed at the language level. The existence of a winning strategy is indeed part of a first-order language—as shown by the introduction of the truth predicate in IF logic. Moreover, this explicit theory of meaning is neutral in relation to classical or constructive logic. However, the proof of the axiom of choice is constructive and its game theoretical interpretation is antirealist after all.

PART IV

Anaphora[90]

This chapter will be dedicated to the second use of IF and GTS that was announced in the introduction: the resolution of anaphora. I mentioned there that, beside the use of IF logic for the foundation of mathematics, Hintikka and his associates think that IF logic and GTS are more suitable for a formalisation of natural languages. One of Hintikka's most cherished example of a language phenomenon that can be treated in a satisfactory manner with IF logic is anaphora.

Hintikka and his associates' work on anaphora, based on GTS, constitutes a landmark in the field and triggered many valuable contributions and discussions. In this section I will compare the GTS approach to the problem of anaphora with a solution for anaphora in the dialogical approach to CTT. The GTS approach has considerable advantages over other existing theories in dealing with anaphora. However, the extension of the dialogical framework discussed in the preceding sections contains both the contentual (first-order) features of CTT and the interactive aspects of GTS.

IV. 1. The GTS approach to anaphora

The challenge is to find a satisfactory account of anaphoric expressions occurring in sentences such as:

1. *If Michael smiles he is happy.*

2. *If a man smiles he is happy.*

3. *Every man that smiles is happy;*

and of more problematic examples such as the famous donkey sentence:

---

[90] This chapter will be published for the most part in (Rahman, Clerbout, Jovanovic, forthcoming).

4. *Every man who owns a donkey beats it.*

The first sentence is not apparently problematic. The Pronoun "he" has a strict interpretation (Michael), so it can be treated as a singular term. The issue is to provide a satisfactory semantic analysis of pronouns "he" or "it" which becomes more challenging when there is an interplay between pronouns and indefinites (such as in 2).

As mentioned previously, Hintikka (1997a) presented the first game-theoretical analysis of anaphora that stressed the distinction between priority and binding scope.[91] According to this analysis, if a quantifier is understood as a logical expression then we are speaking of its *priority scope* in relation to the rest of the sentence; but if it is understood as the antecedent for anaphoric pronoun that appears in the rest of the sentence we are speaking of its *binding scope*. It is a pity, according to the author, that these two different moments are expressed by the same syntactic expression. At first glance, it is appealing to interpret an anaphoric pronoun as a variable available for a quantification. But Hintikka (1997a: 530) contests this view:

> [...] *they do not behave like bound variables. An anaphoric pronoun does not receive their reference by sharing it with the quantifier phrase that is its "head", anymore than a definite description does. An anaphoric pronoun is assigned a reference in a semantical game through a strategic choice of a value from the choice set by one of the players. When the member of the choice set whose selection is a part of the winning strategy of the player in question happens to be introduced to the choice set by a quantifier phrase, that phrase could perhaps be called the head of the pronoun. But, as was pointed out, the origin of the members of the choice set does not matter at all in the semantical rules for anaphoric pronouns.*

The GTS approach to anaphora was further developed by Sandu and his associates.[92] Once again, players' strategies, introduced at the semantic level, are Skolem functions that show a player which disjunct or conjunct or which individual in the model to choose every time it is the player's turn to play. The Skolem form of a formula can always be obtained from the formula in the prenex normal form by systematically replacing every existential quantifier with an appropriate Skolem function, the argument of which is a variable bound by a

---

[91] GTS was used before in the context of the analysis of natural language. Indeed, already in 1985 Hintikka and Kulas, made use of GTS in order to provide semantics of definite descriptions.

[92] Sandu (1997), Sandu, Jacot (2012).

universal quantifier in the scope of which that existential quantifier lays.[93] Now, the idea worked out by Sandu and Jacot (2012) is that indefinites can be represented as Skolem terms. Let us look at the first part of the donkey sentence:

5. *Every man owns a donkey.*

In a game played for this sentence, the Falsifier first chooses an individual that satisfies the predicate of being a man. Then it is the Verifier's task to find a donkey owned by that individual, in order to win the game. The game can be represented as a tree with branches that show all possible outcomes of the game (for any individual chosen by the Falsifier). The Verifier's strategy is a function f that for any individual a, chosen by the Falsifier, gives as a result $f(a)$, that is, a donkey owned by a. Sentence 5 is formalised in the following manner:

6. $\forall x \ (Man(x) \rightarrow Donkey \ (f(x)) \wedge Own(x, f(x)))$.

We can now turn to the anaphoric pronoun in sentence 4. The solution to the problematic anaphora is found with the Skolem term. The pronoun "it" is a copy of the Skolem term in the antecedent. The formalisation of sentence 4 is:

7. $\forall x \ (Man(x) \wedge Donkey \ (f(x)) \wedge Own(x, f(x)) \rightarrow Beats \ (x, f(x)))$.

Sandu and Jacot (2012: 620) claim that Skolem terms are very useful semantic tools for the interpretation of anaphora because they keep track of the entire history of a game. All the variables bound by the quantifiers superior to the indefinites are found as the arguments of each Skolem term. This solution at once combines the quasi-referential view on quantifiers, which is appealing when an anaphoric pronoun appears in a sentence, and the idea of semantic dependency, which is needed both when there is a nesting of indefinites and where there is an interplay of indefinites with quantifiers. In this way one can find a solution to some complicated examples of anaphora.

However, there are some more complicated cases of dependency that involve Henkin's branching quantifiers. In order to deal with those, GTS is combined with IF logic. I presented earlier a typical example of branching quantifiers:

---

[93] See section *1.5.3.*

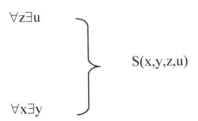

and I explained that this cannot be expressed with a linearly disposed expression of classical first-order language. However this can be done in IF with the following formalisation:

8. $\forall x \forall z\ (\exists y/\forall z)\ (\exists u/\forall x)\ S(x, y, z, u)$

The slash signs indicate that $\exists y$ iz independent of $\forall z$ and that $\exists u$ is independent of $\forall x$. Recall that IF logic captures the precise expressive power of Henkin's branching quantifiers on the first-order level.

A classical example of a natural language sentence that involves branching quantifiers presented by Hintikka (1973: 344) is:

9. *Some relative of each villager and some relative of each townsman hate each other.*

This sentence is expressed by the existential second order sentence:

10. $\exists f\ \exists g\ \forall x\ \forall z\ ((\ Villager(x) \wedge Townsman\ (z)) \rightarrow Relative\ (x, f(x)) \wedge$

Relative (z, g(z)) ∧ Hate (f(x), g(z)))),

which can be translated into first-order IF language as in formalization 8.

According to Hintikka, IF allows us to take into account different patterns of dependency among logical expressions that can appear in a sentence, and it is thus more appropriate for the translation of natural languages. Hintikka (1997a: 523) states the following:

> ...this second-order statement expresses the logical form of the given natural-language sentence. It is equivalent to an IF first-order sentence, which can also be considered as the translation of the given natural language sentence into logical notation.

However, the IF approach is necessarily a model-theoretic one (since IF logic has no proof system). Let me repeat my point: there might be some other reasons forcing us to accept a model theory—these must be discussed independently—but neither the semantic analysis of anaphora nor that of branching quantifiers does so.

IV.2 The dialogical approach to anaphora

In what follows I will give a dialogical account of anaphora, making use of CTT. I will argue that the GTS approach, which puts emphasis on expressing the dependence relations in terms of choices resulting from interaction, is indeed a good way of dealing with anaphora. Moreover, the "outside-in" semantics seems to be much more promising in the analysis of a natural language then the classical "inside-out" approach. However, the dialogical approach provides a first-order solution that does not require any devices other than those of constructive or even classical logic when formulated within a suitably adapted CTT framework. In fact, Hintikka's remark on the *binding scope* touches a crucial point in the semantics of anaphora. In the approach that I am defending, this point involves dependencies between play objects in general and choices for the substitution of instructions (occurring at the object-language level) in particular. This approach is closer to the recent Skolem-term framework developed by Gabriel Sandu and Justine Jacot than to the original analysis of Hintikka, although the dialogical framework can also deal with more complicated cases involving branching quantifiers without making use of IF logic.

In the dialogical approach, pronouns are considered in their anaphoric role. Moreover, in the dialogical framework linked with CTT, the dependence upon a context is taken seriously—that dependence has itself an anaphoric structure. Ranta (1994: 78) introduced the *pronominalisation rule for inferences*, where the dependence of anaphoric pronouns on the context is rendered explicit. "Context" should be understood in the technical sense of CTT as the assumption that the chosen object is of a given type (e.g. the assumption that x is of type A). Thus the pronouns' (and more generally the indexicals') dependence upon a context is understood as a reference to any object of the appropriate type.

For example, Ranta's inferential rules for the pronoun *he* deploys the identity mapping on the set of *man*, while and *she* deploys the identity mapping on the set of *woman* as resulting from the contextual dependence:

a: man

----

he(a): man

Where the rule of a substitution is:

a: man

----

he(a) = a: man[94]

In the previous rule the instance lent by the context is a constant expression, but it can also be a variable—in which case the context would be an open assumption. The general inference rule introduced by Ranta (1994: 80) is the following:

$$a(x_1, ..., x_n) : man \ (x_1 : A_1, ..., x_n : A_n \ (x_1, ..., x_{n-1}))$$

$$he(a(x_1, ..., x_n)) : man \ (x_1 : A_1, ..., x_n : A_n \ (x_1, ..., x_{n-1})),$$

$$he(a(x_1, ..., x_n)) = a(x_1, ..., x_n) : man \ (x_1 : A_1, ..., x_n : A_n \ (x_1, ..., x_{n-1}))$$

Ranta (1994: 80) introduces one more rule for the treatment of pronouns, namely the *sugaring rule*:

$$he(a)(a : A) \ \triangleright \ he: A$$

This rule allows the use of *he* alone in the context A, without further concerning argument a.

Let us now see how these rules work in the dialogical framework. If player X posits that *he (a): man*, then his opponent can challenge this posit by asking him to show that *a* is of the type *man*. Once again, we make use of the instructions, which allows taking variables into the consideration as well.

---

[94] Ranta (1994, p. 78) unites those two rules in one rule with two conclusions.

Posit	Challenge	Defence
X he(I^{pron}): man	**Y** $?_1^{pron}$: man	**X** I^{pron}: man

The resolution of the instruction in I^{pron}: *man* allows the defender to explicitly introduce an identity within the set *man*:

Posit	Challenge	Defence
X I^{pron}: man	**Y** $_I^{pron,man}/?$:	**X** he(I^{pron}) = I^{pron} : man

The sugaring rule is as follows:

Posit	Challenge	Defence
X he(a) = a: man	**Y** $?_{a/he(a)}$	**X** $\varphi[a]$
...		
X $\varphi[he(a)]$		

The sugaring rule allows the challenger to substitute *he(a)* for *a* at any place where *h(a)* occurs, given that the identity *he(a) = a: man* is established and given that the suitable play-object has already been substituted for the instruction I^{pron}. Similar rules can be formulated for *she*. The case of *it* requires more care, since its type might vary from context to context. In any case, some variations are possible for all pronouns (e.g. *she: ship*).

Let us come back to our example of the happy man:

If a man smiles he is happy.

The idea is that in order to obtain an interpretation of the pronoun *he*, we first formalize the first part "*A man smiles*" as:

($\exists x$: man) smiles(x),

and then consider the sentence "*he is happy*" in the context:

> $z: (\exists x: man)\ smiles(x)$.

This analysis yields the following formalization:

$$(\forall z: (\exists x: man)\ smiles(x))\ (happy\ (he\ (L^{\forall}(z)))^{95}$$

In the dialogical framework the left part of the universal given above consists of the set of all men that smile and the right part claims that an object chosen from that set is happy. The left part of the universal is $L^{\forall}(z)$, that is, the set of all men who smile. Its role is to pick up one of the individuals of the set of smiling men (the set $(\exists x: man)\ smiles(x)$). Let us now look at the dialogical game for this sentence. Let us not make use of the instruction I^{pron} now; let us assume that the pronoun has picked an instruction already. Also let us leave out the moves involving the choice of repetition ranks for the sake of simplicity:

	O			**P**	
				$(\forall z: (\exists x: man)\ smiles(x))\ (happy\ (he\ (\ L^{\exists}(L^{\forall}(z)))$	0
1	$L^{\forall}(z):\quad (\exists x:\quad man)\ smiles(x))$	0	1	$R^{\forall}(z): happy\ (he((L^{\forall}(z))$	12
3	$a: (Ex: man)\ smiles(x)$			$L^{\forall}(z)/?$	2
5	$L^{E}(a): man$		3	$?L$	4
7	$R^{E}(a): smiles(L^{E}(a))$		3	$?R$	6
9	$a_1: man$		5	$L^{E}(a)/?$	8
11	$a_2: smiles(a_1)$		7	$R^{E}(a)/?$	10
13	$?L^{\forall}(z)/?\ ...$	12		$R^{\forall}(z): happy\ (he(a_1))$	14
15	$?\ _{a1:\ man}$	14		$a_1: man$	16

---

95 Ranta (1994, p. 79) gives as an exaple "If a man walks he talks".

17	$I^{he,man}/?$	16		$R^\vee(z) : happy\ (he(a_1)= a_1)$	18
19	? a/he(a)	18		$R^\vee(z) : happy\ (a_1)$	20
21	$?R^\vee(z)/$			**P** loses unless he can force **O** to concede that there is a play object b for *happy (a₁)*, such that it allows **P** to choose b for $R^\vee(z)$ while responding to the challenge of move 21 on move 12	

Description:

Move 0: **P** states the thesis.

Move 1: **O** challenges the universal by positing an arbitrary man that smiles, that is z: $(\exists x: man)\ smiles(x)$.

Move 2: **P** counterattacks by asking who that man is.

Move 3: **O** responds by choosing some play object.

Move 4: Since a is an existential it is constituted by two parts: **P** starts by asking for its left part.

Move 5: **O** answers that $L(a)$ is a man.

Move 6: **P** challenges now the right part of the existential.

Move 7: **O** responds to the attack.

Move 8: **P** asks **O** to resolve the instruction occurring in the expression brought forward in move 5

Move 9: **O** responds by choosing a_1.

Move 10: **P** asks **O** to resolve the instruction occurring in the expression brought forward in move 7.

Move 11: **O** responds by choosing a_2..

Move 12: **P** answers now the challenge of move 1

*Move 13: **O** asks **P** to resolve the instruction occurring in the expression brought forward in move 12.*

*Move 14: **P** chooses a_l*

*Move 15: **O** challenges the pronoun he*

*Move 16: **P** can answer a_l: man, since **O** conceded it before (namely in move 9)*

*Move 17: **O** forces **P** to bring forward the identity underlying the pronoun he*

*Move 18: **P** brings forward the required identity*

*Move 19: **O** forces **P** to make use of the identity brought forward in move 18 and apply it to drop the pronoun occurring in 14*

*Move 20: **P** drops the pronoun and this yields $R^\forall(z)$: happy (a_l)*

*Mover 21: O asks to resolve the instruction occurring in the last move. Since it is an elementary expression and **O** did not concede it before **P** cannot has no move to play and lost the play.*

    P has a winning strategy if in a given context a_l is a man who smiles and is happy and stands for every choice of man that **O** can make. Of course, this sentence is not valid. One could develop a material dialogue by introducing the initial concessions of the Opponent (the premises) and thus checking whether there is a winning strategy. The winning strategy for the Proponent then amounts to an inference from the materially-given premises. Remember that to design a material dialog the model is not introduced in the way Hintikka introduces it, but is rather specified as the number of initial concessions by Opponent. The dialogical approach is not a model-theoretic one.

    Let us now consider the famous donkey sentence. The analysis made by Sundholm (1986) constitutes a landmark in the application of CTT to natural language. In order to keep the focus on the interdependence of choices, let us skip the pronouns *he* and *it* and replace them with the corresponding instructions.

    Every man who owns a donkey beats it.

As in the example above, let us first formalize the first part of the sentence "man who owns a donkey" and then consider that sentence in context, such that we obtain:

$z: (\exists x: man) (\exists y: donkey)(x\ owns\ y).$

Since the existential is in fact the set of which z is an element, it is more convenient to use the set-separation notation:

$z: \{x: man \mid (\exists y: donkey)(x\ owns\ y)\}.$

Take the left part of z to pick up a man (that owns a donkey). The right part of z is the owned donkey (that is beaten). Put together it yields the following formulation:

$p: (\forall z: \{x: man \mid (\exists y: donkey)(x\ owns\ y)\}) (L^{\{...\}}(z)\ beats\ L^{\exists}(R^{\{...\}}(z)))$

Or more briefly:

$p: (\forall z: \{x: M \mid (\exists y: D)\ Oxy\})B\ (L^{\{...\}}(z), L^{\exists}(R^{\{...\}}(z)))$

Let us now run the play for a material dialog. It is convenient to ignore the choice of repetition ranks once again. Since this is a material dialogue, the formation rules indicate how the sets are composed. It is also known that m is a man, that d is a donkey, and that p is a play object for the proposition $m\ owns\ d$.

	O			**P**	
I	*M: set*				
II	*D: set*				
III	*Oxy: set (x:M, y:D)*				
IV	*Bx,y: set*				
V	!p: $(\forall z: \{x: M \mid (\exists y: D)$ Ox y$\})B$ $(L^{\{...\}}(z), L^{\exists}(R^{\{...\}}(z)))$				
VI	!m: M				

VII	$!d: D$				
VIII	$!p': Omd$				
				$!B(m,d)$	0
1	$n: = ...$			$m: = ...$	2
3	$? play\text{-} object$	(o)		$!q: B(m,d)$	30
25	$!R^\forall(p):B(L^{\{...\}}(z), L^\exists(R^{\{...\}}(z)))$		V	$!L^\forall(p): \{x : M \mid (\exists y: D)\ Oxy\}$	4
5	$L^\forall(p)/? ...$	(4)		$!z: \{x : M \mid (\exists y: D)\ Oxy\}$	6
7	$? L$	(6)		$! L^{\{...\}}(z): M$	8
9	$L^{\{...\}}(z)?$	(8)		$!m: M$	10
11	$? R$	(6)		$! R^{\{...\}}(z): (\exists y: D)\ Omy$	12
13	$R^{\{...\}}(z)?$	(12)		$L^\exists(R^{\{...\}}(z)), R^\exists(R^{\{...\}}(z)): (\exists y: D)\ Omy$	14
15	$L^\exists(R^{\{...\}}(z))/?, R^\exists(R^{\{...\}}(z))/?$	(14)		$!(d,p'): (\exists y: D)\ Omy$	16
17	$? L$	(16)		$! L^\exists(d,p'): D$	18

19	$L^{\exists}(R^{\{...\}}(z))/?,$	(18)			!d: D	20
21	? R	(16)			! $R^{\exists}(d,p')$: Omd	22
23	$R^{\exists}(R^{\{...\}}(z))/?$	(22)			!p': Omd	24
27	!$R^{\forall}(p)$:B(m,d)			(25)	$L^{\{...\}}(z)/m,\ L^{\exists}(R^{\{...\}}(z))/d$	26
	!q: B(m,d)			(27)	$R^{\forall}(p)/?$	28

Description:

Moves I- VIII: *These moves are **O**'s initial concessions. Moves I- IV deal with formation of expressions. After that the Opponent concedes the donkey sentence and atomic expressions related to the sets M, D and Oxy.*

Moves 0- 3: *The Proponent posits the thesis. The players chose their repetition ranks in moves 1 and 2. The actual value they chose does not really matter for the point we want to illustrate here, thus we simply assume that they are enough for this play and live them unspecified. Now, when **P** posited the thesis he did not specified the play- objects so **O** asks for it in move 3.*

Move 4: ***P** chooses to launch a counter- attack by challenging the donkey sentence which **O** conceded at V. The rules allow him to answer directly to the challenge, but then he would not be able to win.*

Move 5 – 24: *The dialog then proceeds in a straightforward way with respect to the rules introduced earlier. More precisely, this dialog displays the case where **O** chooses to challenge **P**'s posits as much as she can before answering **P**'s challenge 4.*

Notice that the Opponent cannot challenge the Proponent's atomic expressions posited at moves 10, 20 and 24: since O made the same posits in her initial concessions VI- VIII, the modified formal rule SR3 forbids her to challenge them.

*Move 25: When there is nothing left for her to challenge, **O** comes back to the last unanswered challenge by **P** which was move 4 and make the relevant defence according to the particle rule for universal quantification.*

Moves 26- 27: The resolution for instructions $L^{\{...\}}(z)$ and $L^{\exists}(R^{\{...\}}(z))$ has been carried out during the dialog with moves 9- 10 and 23- 24. Thus the Proponent can use the established substitution to challenge move 25 according to the structural rule SR4.2. The Opponent defends by performing the requested substitutions.

*Moves 28- 30: The Proponent then asks the play - objects for which the instruction $R^{\forall}(z)$ stands. When she answers, the Opponent posits exactly what **P** needs to defend against **O**'s challenge 3. Notice that at this point this is the last unanswered challenge by **O**, therefore **P** is allowed to answer it in accordance to the structural rule SR1i. He does so with his move 30.*

*Since **O** made the same posit the rule SR3 forbids her to challenge it. She then has no further possible move, and the Proponent wins this dialog.*

Recently, a paper[96] on anaphora appeared by Darryl McAdams and Jonathan Sterling where they offer a solution very similar to one developed here, making use of the *Require* rule. They worked out a computational and proof-theoretical justification of this rule. Let me just mention that this kind of rule actually has the most natural justification in game-theoretical framework, where the questions and answers are already an integral part of meaning explanation.

VI. 3. Branching quantifiers

In the dialogical framework more difficult examples involving branching quantifiers can also be handled in a satisfactory manner. Let me repeat once again the example:

---

[96] McAdams, Sterling (2015).

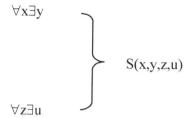

The Skolem form of this sentence is:

$$\exists f \exists g \forall x \forall z\; S(x, f(x), z, g(z))$$

Sundholm (2013a) gives the CTT analysis of branching quantifiers with the following presentation:

$$(\exists f \in (\Pi x \in D)D)(\exists g \in(\Pi x \in D)D)(\forall x \in D)(\forall z \in D)A[ap(f, x)/y, ap(g, z)/u].$$

the explanation is:

> *The quantifiers in this Skolemziation are with respect to elements f and g of the function set ($\Pi x \in D)D$, but do not quantify over a function type. The function sets, like all sets of Constructive Type Theory, are predicative and generated from below. Their elements correspond to Russell's entities of lowest level, that is, to individuals.*

The dialogical development is straightforward. Recall that the application of a function follows this rule:

Posit	Challenge	Defence
$X\,!\,p : \varphi\,[f(k_1)]$	$Y\,f(k_1)/?$	$X\;p : \varphi\,[k_2\,/\,f(k_1)]$ $<\varphi\,[f(k_1)] = \varphi\,[k_2/\,f(k_1)] : set]>$

Moreover, in the dialogical framework sentences involving branching quantifiers can be handled on the first-order level in such a way that all objections against branching quantifiers can be now answered in a satisfactory manner. The objections relying on the fact that the semantics of branching quantification necessarily involves quantification over functions was dismissed in *I.6.2.* using arguments from Sundholm. But let us recall Patton's objection from the same section: Patton says that branching quantification should not actually be considered quantification at all because the necessary condition for being a quantifier is having instances. Notice that the treatment of branching quantifiers in the dialogical approach allows having instances! There are very good reasons to state that branching quantifiers are of the first-order level after all—and this interpretation is supported by the dialogical framework.

Finally, recall Ranta's remark, mentioned in *I.7.*, that Hintikka's argument against constructivist proof-theoretical semantics comes down to its alleged incapacity in providing a satisfactory interpretation of branching quantification. This is now clearly dismissed by the account of branching quantifiers offered in the dialogical approach linked with CTT: this interpretation of branching quantifiers does not push us toward model theory after all.

IV. 4. Final remarks

As already mentioned, Hintikka's remark on *binding scope* touches a crucial point in the semantics of anaphora, which naturally leads to the game-theoretical interpretation. In the dialogical game-theoretical framework this point involves dependencies between play objects in general and the choices for a substitution of instructions in particular. It was necessary for a framework such as that of CTT to make this point about dependencies explicit. The dialogical approach implements these dependencies within a game-theoretical analysis. Indeed, without such an approach choice dependencies cannot be expressed in the object-language. Besides, in dialogues these dependencies can be expressed without the possibility of some non- desirable phenomena such as *signalling*. This is one more reason to favour the dialogical over the GTS approach.

One other advantage of the dialogical approach is that the meaning of anaphoric expressions is obtained at the play level and not through the existence of a winning strategy for a player, as in GTS. That is an advantage because it shows how one can understand the meaning of an anaphoric pronoun without knowing how to win the game: it is enough that one understands all the steps the Proponent is committed to in a dialogical game.

Conclusions

It is time to briefly summarise the main results of this book. I have focused on Hintikka's Independence Friendly Logic and the main uses he makes of it: first as the medium for mathematical reasoning and second as a means of formalising and analysing natural languages. I presented it in some detail, trying to stay faithful to Hintikka's own line of thought. I discussed the properties of IF logic as well as the advantages of this approach: the possibility of taking account of dependency and independency relations among variables; the GTS account of two different notions of the scope of quantifiers; the "outside-in" direction for approaching meaning, which turns out to be advantageous over the traditional "inside-out" approach; the usefulness of game-theoretical reasoning in mathematics; the expressiveness of IF language, which allows for formulating branching quantifiers at the first-order level, as well as for defining the truth predicate in the language itself. I defended Hintikka's standing of the first-order character of IF logic against some objections on this point, given a particular interpretation of IF.

However, I have also brought to the fore some of the weaker sides of Hintikka's proposal: first and foremost, the lack of a full axiomatisation for IF logic and second, the problem of signalling, a phenomenon related to the possibility of imperfect information in a game. As for the first, I argued that while it is certainly true that there are many other domains in which validity is not our main interest (like when it comes to the analysis of natural languages), it is hard to argue the same when it comes to mathematical reasoning. Moreover, without entering into a deep discussion concerning the advantages of a proof-conditional semantics over a model-theoretic one, it has been argued that neither the semantics of branching quantifiers not that of anaphora really force us to adopt a model-theoretic point of view. Hintikka's argument against a proof-conditional approach to meaning—that it cannot deliver a satisfactory account of branching quantifiers—has been refuted.

In the meantime, I presented another game-theoretically oriented semantics, namely dialogical logic linked with Martin-Löf's Constructive Type Theory. I showed that in this framework dependency and independency relations can be accounted for, but without using more means than constructive logic and the dialogical approach to meaning have to offer. Thus, I used it first to analyse Hintikka's take on the axiom of choice and then to analyse the GTS account of anaphora.

I contrasted Martin-Löf's analysis of the axiom of choice with Hintikka's standing on this axiom. Hintikka claims that GTS for IF logic justifies Zermelo's axiom of choice at the first-order level in a way perfectly acceptable to constructivists. I used Martin-Löf's result to develop several important points. I argued that there is a reading of Hintikka's preferred version of the axiom of choice that is indeed acceptable for constructivists and that under that reading its meaning does not involve a higher-order logic. However, to render this version acceptable for constructivists, one has to adopt an intensional take on functions, while extensionality is an essential part of the classical understanding of Zermelo's axiom and this is the real reason why constructivists reject it. Moreover, I argued that the restriction to recursive functions that Hintikka proposes in order to offer a constructive reading of GTS is not actually the most relevant. What is needed is the intensionality of the choice function. More precisely, functions are understood intensionally in the sense that they are given as rules of correspondence by which the value of a function is constructed out of the argument. This is clearly the case in the dialogical approach that I have defended here: functions are conceived as rules of interaction.

I used these results to arrive at a further point: the dependence and independence features that motivate IF logic can be formulated within the CTT frame, which offers several advantages. One can express all the dependence patterns but without paying the price of having a system that is neither axiomatizable nor has an underlying theory of inference.

Let me underline one specific point I have defended here: a game-theoretical framework promises to be very fruitful in the context of the foundations of mathematics, and Hintikka is indeed justified in appealing to such a framework. However, I argued that a framework such as that of dialogical logic shows itself to be more convenient for this purpose. Moreover, recent developments in dialogical logic show that the CTT approach to meaning in general and to the axiom of choice in particular is very natural to game theoretical approaches where meta-logical features are explicitly displayed at the object-language level.

One of the fundamental features of the dialogical approach to CTT is that all the actions that constitute meaning are rendered explicit in the object-language. The game theoretical reading of the axiom of choice stresses one of the more salient characteristics of CTT language: the judgment that a proposition is true can be expressed at the language level. The existence of a winning strategy is part of a first-order language indeed—as pursed by the introduction of a truth predicate in IF logic. Moreover, this explicit theory of meaning is neutral in relation to classical or constructive logic. However, while the proof of the axiom of choice is constructive its game theoretical interpretation is antirealist after all.

In the last section, I analysed the GTS account of anaphora. I argued that the GTS approach has considerable advantages over other existing theories in dealing with anaphora. Yet again, CTT in the dialogical framework contains both the contentual first-order features of CTT and the interactive aspects of GTS.

Hintikka's remark on *binding scope* touches the crucial point in the semantics of anaphora that naturally leads to the game-theoretical interpretation. However, I argued that this point involves dependencies between play objects in general and the choices for a substitution of instructions in particular. It was necessary that a framework such as that of CTT make the point of these dependencies explicit. The dialogical approach implements these dependencies within a game-theoretical analysis. Indeed, without such an approach choice dependencies cannot be expressed at the object-language level. Besides, in the dialogues those dependencies can be expressed without the possibility of some non-desirable phenomena such as *signalling*. This is one more reason for favouring the dialogical over the GTS approach.

One other advantage of the dialogical approach is that the meaning of anaphoric expressions is obtained at the play level and not through the existence of a winning strategy for a player, as in GTS. This is an advantage because it shows how we can understand the meaning of an anaphoric pronoun without knowing how to win the game: it is enough that one understands all the steps to which the Proponent is committed in a dialogical game.

Finally, I would like to make one more remark concerning the dialogical approach to CTT. The CTT dialogue proves to be a very powerful tool, both for the analysis of a natural language and potentially for the foundations of mathematics. Its subtle approach to meaning, as well as its fully interpreted language, offers a number of possibilities for future application. CTT has been already shown to be very fruitful for foundational purposes in mathematics. In the dialogical game-theoretical framework it can be even more interesting for this purpose. It is a challenging idea for future work in this domain.

Appendix

Dialogues for first-order logic

Let me first define the vocabulary for first-order logic in the same way as in section *I.1.1.*. The validity of a first-order formula is proved through a dialogical game for the formula between two players: the *Proponent*, who defends the formula, and the *Opponent*, who tries to falsify it. As such, one more set of additional signs is needed:

$S_1 = \{x, y, z... \}$ — a countable set of variables.

$S_2 = \{\neg, \wedge, \vee, \exists, \forall \}$ — a set of logical symbols for negation, conjunction, disjunction, existential and universal quantifier, respectively.

$S_3 = \{ R_n$: n is an element of the set of natural numbers N $\}$ — the set of relational symbols. Every such relational symbol R has a natural number n that indicates its arity. Unary relational symbols are *predicates.*

$S_4 = \{ f_m$: m is an element of the set of natural numbers N $\}$ — the set of functional symbols. Every such functional symbol f has a natural number m that indicates its arity. Nullary function symbols are *constant symbols,* which will be defined in a separate set.

$S_5 = \{c_1, c_2 ... \}$ — the set of constant symbols.

$S_6 = \{ ( ) , = \}$ — the set of additional symbols.

$S_7 = \{P, Q\}$, where P and O correspond to players in the dialogical game (Proponent and Opponent, respectively).

$S_8 = \{?, !\}$ is another set of additional signs where "?" and "!" are *force symbols,* indicating that the move in the game is an *attack* or a *defence.*

$S_9 = \{ X, Y \}$ is a set of *players* in a dialogical game for a formula that can assume the roles of Opponent or Proponent in the game.

Let me present a few definitions:

Definition 1:

Dialogically signed expressions are strings <P-!-φ>; <O-!-φ>; <P-? -φ> and <O-?-φ> indicating who is making a move for an expression φ and if it is an attack or a defence.

Definition 2:

> *The dialogical game* is a sequence of dialogically signed expressions.

Definition 3:

> *The play* is a sequence of dialogical games.

Dialogical games can be seen as a special case of plays: all dialogical games are plays but not vice versa. As I will show later, a certain rule[97] for dialogues allows for the splitting of a game into two branches. Each of the branches will be a separate dialogical game, but taken together they will form a play.

Definition 4:

> A *dialogue* is a set of plays.

Definition 5:

> The thesis of the dialog D(φ) is φ, that is, the formula asserted by the initial Proponent in the dialogue.

Definition 6:

> I will denote Δ[n] a member of the sequence Δ of the dialogue[98] that has a position n.

As mentioned above, a dialogical game for a first-order formula is played by two players: the Proponent, who tries to defend the formula, and the Opponent, who tries to challenge it. The game is *finite*, i.e., after a finite number of steps one of the players wins while other loses. The course of the game is determined with two groups of rules: *particle rules* and *structural rules*.

---

[97] See Branching rule for dialogical games.
[98] which is a dialogically signed expression.

Particle rules

Particle rules for dialogical games provide the so-called *local semantics*. They show how the game is played locally for every logical constant in the formula. These rules indicate how every logical constant can be attacked and defended and whose turn is to play in every step of the game. Particle rules therefore yield local semantics independently of the context in which the argumentation takes place. Context-dependence enters into the game with structural rules.

Atomic formulas have no particle rule corresponding to them in formal dialogues.

Assertion	Attack	Defence
X-!-$\varphi \wedge \psi$	Y-?-L or Y-?-R The choice is made by the player in the role of Opponent	X-!-φ or X-!-ψ respectively
X-!-$\varphi \vee \psi$	Y-?-$\vee$	Y-?-L or Y-?-R The choice is made by the player in the role of Proponent
X-!-$\varphi \rightarrow \psi$	Y-!-φ	X-!-ψ Defence is not possible.

X-!-¬ φ	Y-!-φ	The only possible move is counterattack.
X-!-∀x φ	Y-?-∀x / a	X-!-φ [x/a]
	Y chooses any a available to him	for a chosen by Y
X-!-∃x φ	Y-?-∃	X-!-φ [x/a]
		X chooses any a available to him

"φ [x/a]" in the diagram means that every occurrence of x in the formula φ is substituted by a. "L" and "R" stand for the *left* and *right* part of the conjunction or disjunction.

Particle rules describe the course of the dialogical game, or how we can pass from one *state of the dialogue* to another.

Definition 7:

Let φ be a first-order formula, and {a₀, a₁...} a countable set of individual constants. Then the *state of dialog* S is defined by specification of the following components:

1) ψ—a subformula of φ

2) <X- Y-e>—a dialogically signed expression where X stands for either P or O, i.e., the roles of players in the game; Y stands for "?" or "!", i.e. it indicates whether the move is an attack or a defence; and e = ψ.

3) σ: Free [ψ] → { a₀, a₁...}—a function from free variables in B to individual constants.

4) ρ: {P, O} → {?, !}—a function assigning the roles to players.

If we take that we are in the state S₁ = <ψ₁, X₁, Y₁, e₁, σ₁> of the dialog D(φ), then particle rules show how the dialog can evolve from the state S₁ to the state $S_2 = \langle \psi_2, X_2, Y_2, e_2, \sigma_2 \rangle$. S_2 is reached when the player X_1 makes a move with the force Y_1 in accordance with the appropriate particle rule. Depending on the Y_1 the move is either an attack (?) or a defence (!).

Definition 8:

> A pair formed of an attack and a defence are a *round* in a dialogic game. The round is opened by an attack and closed with a defence.

Finally, the particle rules are as follows:

PR∧ (particle rule of conjunction):

If $e = \psi$, where ψ is the subformula of φ of the form $\alpha \wedge \beta$, and $Y = !$, then player Y choses to attack the left or the right part of the conjunction, so the attack is $\langle Y\text{-?-}L\rangle$ or $\langle Y\text{-?-}R\rangle$. The state of dialog S_2 is either $\langle \alpha, X, !, \alpha, \sigma \rangle$ or $\langle \beta, X, !, \beta, \sigma \rangle$, respectively.

Intuitively, for the Opponent to falsify the conjunction it is enough to show that one of the conjuncts can be refuted. Thus, the Opponent decides which conjunct is easier to challenge. In response, the Proponent of the conjunction must be able to defend the challenged conjunct.

PR∨ (the particle rule of disjunction):

If $e = \psi$, where ψ is the subformula of φ of the form $\alpha \vee \beta$, and $Y = !$, then the attack consists in the Opponent's question ?-∨, and the state of dialog S_2 is either

$\langle \alpha, X, !, \alpha, \sigma \rangle$ or $\langle \beta, X, !, \beta, \sigma \rangle$, depending on the choice of the Proponent who is defending the disjunction.

Intuitively, for the disjunction to be defended it is enough to show that one of the disjuncts can be defended, so the Opponent is challenging the Proponent to defend one. It is then up to the Proponent to choose which disjunct she wants to defend.

PR→ (the particle rule of implication):

If $e = \psi$, where ψ is the subformula of φ of the form $\alpha \rightarrow \beta$, and $Y = !$, the attack of the Opponent in S_2 consists in defending the antecedent of the conditional,

$S_2 = <\alpha, Y, !, \alpha, \sigma>$ Then in the state S_3 it is the Proponent who must defend the consequent, so $S_3 = <\beta, X, !, \beta, \sigma>$, or else the Proponent can challenge the Opponent's defence of α to yield the state S_4 of the dialog, $S_4 = <\alpha, X, ?,..., \sigma>$.[99]

Intuitively, to show that the implication is valid the Proponent must be able either to defend both the antecedent and the consequent to show that antecedent can be refuted. So the attack of the Opponent consists in defending the antecedent, thus challenging the Proponent to show that he can defend the consequent. The Proponent can do this or he can try to refute the antecedent by attacking the antecedent in the next stage of the dialogue.

PR$\neg$ (particle rule of negation):

If $e = \psi$, where ψ is the subformula of φ of the form $\neg\,\alpha$, and $Y = !$, then the attack of the Opponent consists in his defending α in stage S_2 of the dialogue.

$S_2 = <\alpha, Y, !, \alpha, \sigma>$. The Proponent does not have a proper defence against this attack, but he can counterattack formula α in the next stage of a dialog $S_3 = <\alpha, X, ?,..., \sigma>$.[100]

The rule of negation amounts to switching the roles of the players in the game. When a player challenges the negation he charges himself with defending the non-negated formula. The only response from the part of a defender of the initial negated formula is to try to refute the formula without negation.

PR$\forall$ (the particle rule of universal quantifier):

If $e = \psi$, where ψ is the subformula of φ of the form $\forall x\ \alpha x$, and $Y = !$, the attack by the Opponent is the question ? $\forall x/a_i$ where a_i is chosen by him. In his response the Proponent must defend the formula where x is substituted by a_i, so the state of dialog S_2 is

$S_2 = <\alpha x, X, !, \alpha\ x, \sigma[x/a_i]>$. As usual, $\sigma[x/a_i]$ is the function replacing all the occurrences of x bound by the universal quantifier with a_i.

This explanation is rather obvious. To defend a formula $\forall x\ \alpha x$ the Proponent must be able to defend αx for any individual constant that can be substituted for x, so it is up to the Opponent to choose the constant. The Proponent has to defend the new formula with the constant substituted for x.

---

[99] In S_4, α is the formula in question and fourth component is determined by a logical form of α i.e. the relevant logical constant in α is attacked.
[100] See the footnote 6.

PR∃ (particle rule of existential quantifier):

If $e = \psi$, where ψ is the subformula of φ of the form $\exists x\ \alpha x$, and $Y = !$, the attack by the the Opponent is the question ?- $\exists x$ and the state $S_2 = \ <\alpha x,\ X,\ !,\ \alpha x,\ \sigma[x/a_i]>$, only this time it is the Proponent who chooses the individual constant for substitution.

The rule for the existential quantifier is the dual of the previous rule. To defend formula $\exists x\ \alpha x$ it is enough for the Proponent to show that αx can be defended for one individual constant substituted for x, so it is up to the Proponent to choose the individual constant.

Structural rules

Structural rules organise the entire dialog. The meaning of a formula is determined from the *outside in*, such that the dialogue starts with an entire initial formula, the thesis, defended by the initial Proponent. Through the rest of the dialogue the Proponent tries to give argumentation in favour of the thesis, so to speak, while the Opponent challenges it. That is done through *moves* made by players in accordance with particle rules.

Formal dialogues serve for testing the logical validity of a formula. The formula is proven to be valid if the Proponent succeeds in defending it against the attack of the Opponent, that is, if there is a winning strategy for the Proponent. This notion of validity captures the standard notion of validity as *the truth in every model*.

A dialogue can be given as a tree representation. The root of the tree is the theses together with a set of other premises (that can be empty). The branches are sequences of dialogically signed expressions generated by the course of the dialog. Branches can only be made by the Opponent's choices—by her attack against a conjunction, her defence of a disjunction, or her reaction against an attack on implication. The moves of Proponent cannot generate a branch.

I shall proceed now with a presentation of structural rules. Let the language be L and φ a first-order formula. The thesis of the dialog $D(\varphi)$ is φ. Recall that we denote $\Delta[n]$ a member of the sequence Δ of the dialogue that has the position n.

SR 0 (Starting rule):

a) The first sequence of the dialog, $\Delta[0]$, is the thesis φ stated by the Proponent together with a set of other initial premises, stated by the Opponent. Thus, we will have dialogically signed expressions $<P\text{-}!\text{-}\varphi>$, $(<O\text{-}!\text{-}h_1>...<O\text{-}!\text{-}h_n>)$ for the set of initial hypotheses $\{h_1...h_n\}$. When proving the validity of a first-order formula, the set of additional premises is empty.

b) After the theses is stated in $\Delta[0]$ the players make moves one after another. Every move is a reaction to an earlier move by the adversary. At even positions it is the Proponent who makes the move and in odd positions it is the Opponent. Therefore, in positions $\Delta[2n]$ the sequence Δ will be of the form $<P\text{-}Y\text{-}\psi>$, where ψ is a subformula of φ and $Y \in \{?, !\}$, while in positions $\Delta[2n+1]$ the sequence Δ will be of the form $<O\text{-}Y\text{-}\psi>$.

SR 1.C (Classical round closing rule):

A player can attack any complex formula asserted by her opponent within the appropriate particle rule and she can defend herself against any attack. If needed, she is allowed to repeat her earlier defences.

SR 1.I (Intuitionistic round closing rule):

A player can attack any complex formula asserted by her opponent within the appropriate particle rule and she can defend herself against *the last attack of her opponent that has not yet been answered.* In other words, repeating earlier defences is not allowed and only the latest attack can be answered if it has not already been answered. A defence can be delayed as long as it is possible to attack the opponent instead.

SR 2 (Branching rule for dialogical games):

As previously mentioned, only certain moves by the Opponent can produce branches in the game tree, namely, his attack on a conjunction, his defence of a disjunction, and his reaction to the Proponent's attack against an implication. In those cases sequence Δ splits into two sequences of the game Δ_1 and Δ_2. No move of the Proponent can create branching, and nor can the Opponent's moves against quantifiers. Each branch of a game is a separate dialogical game.

Let $D(\varphi)$ be a dialogue for a first-order formula φ and let Δ be such that it is the Opponent's turn to move.[101]

---

[101] The number of the position of Δ in the game is odd.

1) If m is a natural number such that $m < n$, $\Delta[m] = <O\text{-}!\text{-}\alpha \vee \beta>$, and $\Delta[n] = <P\text{-}?\text{-}\vee>$, then O can split Δ in two sequences, Δ_1 and Δ_2: $\Delta_1[n+1] = <O\text{-}!\text{-}\alpha>$; $\Delta_2[n+1] = <O\text{-}!\text{-}\beta>$.

2) If $\Delta[n] = <P\text{-}!\text{-}\alpha \wedge \beta>$, then O can split Δ in two sequences Δ_1 and Δ_2: $\Delta_1[n+1] = <O\text{-}?\text{-}L>$; $\Delta_2[n+1] = <O\text{-}?\text{-}R>$.

3) If m is a natural number such that $m < n$, $\Delta[m] = <O\text{-}!\text{-}\alpha \rightarrow \beta>$, and $\Delta[n] = <P\text{-}!\text{-}\alpha>$, then O can split Δ into two sequences, Δ_1 and Δ_2: $\Delta_1[n+1] = <O\text{-}?\text{-}...>$[102]; $\Delta_2[n+1] = <O\text{-}!\text{-}\beta>$.

SR 3 (Shifting rule):

By the application of previous branching rules the play of a dialog can split into two dialogical games triggered by the Opponent's choice. The shifting rule says that if the Opponent has chosen to split Δ into two sequences, Δ_1 and Δ_2, and further has lost play Δ_1, he can come back and try to win play Δ_2. Notice that the Opponent cannot switch to the alternative play Δ_2 unless he has already lost play Δ_1. Let me present an example:

$\Delta[n] = <O\text{-}!\text{-}\alpha \vee \beta>$

$\Delta[n+1] = <P\text{-}?\text{-}\vee>$

$\Delta[n+2] = <O\text{-}!\text{-}\alpha>$.

Suppose that the Opponent has lost this game, i.e. he cannot succeed in defending α. The shifting rule allows him to come back and defend the disjunct β this time.

$\Delta'[n] = <O\text{-}!\text{-}\alpha \vee \beta>$

$\Delta'[n+1] = <P\text{-} ?\text{-}\vee>$

$\Delta'[n+2] = <O\text{-}!\text{-}\alpha>$.

The difference between a *dialogical game* and a *play* can now be explained. Both sequences ($\Delta[n]$, $\Delta[n+1]$, $\Delta[n+2]$) and ($\Delta'[n]$, $\Delta'[n+1]$, $\Delta'[n+2]$) are *plays that are dialogical games*. Put together the sequence ($\Delta[n]$, $\Delta[n+1]$, $\Delta[n+2]$, $\Delta'[n]$, $\Delta'[n+1]$, $\Delta'[n+2]$) is *a play that is not a dialogical game*. We can say that the play is the order of the dialogical games in the dialogue.

---

[102] α is the formula in question and third component is determined by the logical form of α i.e. the relevant logical constant in α is attacked.

SR 4a (Winning rule for dialogical games):

A dialogical game is *closed* if there is an atomic formula asserted in two positions, once by player X and once by player Y.

Let $D(\phi)$ be the dialogue for a first-order formula ϕ and let ψ be the subformula of ϕ such that ψ is atomic. Let $\{\Delta[n_1]...\Delta[n_i]\}$ be the set of sequences that form the dialogical game. The dialogical game is closed if there are two sequences $\Delta[n_j], \Delta[n_k] \in \{\Delta[n_1]...\Delta[n_i]\}$ such that $\Delta[n_j] = \ <O\text{-}!\text{-}\psi>$ and $\Delta[n_k] = <P\text{-}!\text{-}\psi>$.

A dialogical game is *opened* if it is not closed.

A dialogical game is *finished* either if it is closed or if it is open but the rules do not allow any more moves. Finished dialogical games cannot be further continued.

If a dialogical game finishes closed the initial Proponent is the winner. If the game finishes open the Proponent loses.

SR 4b (Winning rule of plays):

A play is *closed* if all the dialogical games in it are closed. If the play consists of the games $\Delta_1...\Delta_n$, Δ, the play is *finished* if all games $\Delta_1...\Delta_n$ are closed and Δ is finished.

If the play is closed the Proponent wins. Otherwise, she loses.

SR 5 (Formal use of atomic formulas):

a) The Proponent is not allowed to introduce atomic formulas. Any such formula must be asserted first by the Opponent.

b) Atomic formulas cannot be attacked.

The explanation of this rule is rather simple. The Proponent wins the game if there is an atomic formula stated both by himself and by the Opponent; thus, it is not in his favour to introduce new atomic formulas. On the other hand, from the Opponent's perspective, the goal is not to have an atomic formula repeated, so he will try to introduce as many atomic formulas as possible.

Before I present the last structural rule I shall give some more definitions.

Definition 9:

The strict repetition of an attack is:

a) an attack against some assertion of the opponent in a game, even if that assertion has already been attacked with the same dialogically signed expression; or

b) an attack against a universal quantifier stated by the opponent with a new constant, even if the universal quantifier has already been attacked with a constant that was new at the moment of the attack.

From a careful reading of a) it should be clear that attacks against a conjunction consisting of moves <Y-?-L> and <Y-?-R> do not count as a strict repetition in the sense described in the definition.

Definition 10:

> *The strict repetition of a defence* is:
>
> a) a defence against some attack by the opponent in a game, even if that attack has already been answered with the same dialogically signed expression; or
>
> b) a defence of an existential quantifier attacked by the opponent in a game with a new constant, even if the existential quantifier has already been defended with a constant that was new in the moment of that attack.

Again, it should be noted that in the case of a defence of a disjunction, moves <Y-?-L> and <Y-?-R> do not count as a strict repetition in the sense described in the definition. Finally, the last rule is:

SR 6.C (Classical "no delaying tactics" rule)

It is forbidden to use strict repetition in a game. This version of the rule is for classical first-order logic and should obviously only be used with the SR 1.C rule.

SR 6.I (Intuitionistic "no delaying tactics" rule):

It is forbidden to use strict repetition in a game, except in the case that the Opponent has introduced a new atomic formula that can now be used by the Proponent. Then Proponent is allowed to perform a strict repetition of her move.

Of course, this rule is appropriate for intuitionistic first-order logic and should only be used with the SR 1.I rule.

Definition 11:

The first-order sentence φ is *classically valid* if all the plays of the classical dialog D(φ) are closed.

Definition 12:

The first-order sentence φ is *intuitionistically valid* if all the plays of the intuitionistic dialog D(φ) are closed.

The dialogical concept of validity has been proved to coincide with standard concepts of validity for classical and intuitionistic first-order logic. The first such proof was provided by Kuno Lorenz.[103] It has since been developed by number of authors, such as Stegmüller, Haas, Felscher, Krabbe, and Rahman, who established the equivalence between other systems of proofs, such as sequent calculi and the tableaux system, and the dialogues.

Let me present two simple examples of classical and intuitionistic dialogues to illustrate how the game works.

Example:

Let us observe a classical dialogue for the propositional formula p∨¬p, where p is an atomic sentence.

	Opponent		Proponent	
			p∨¬p	0
1	?-∨	0	¬p	2
3	p	2		
(1)	(?-∨)	(0)	p	4

---

[103] The proof can be found in collection of papers by Lorenzen and Lorenz (1978).

In the columns for each player we find the numbers of moves in the dialogue. If the move counts as an attack in the innermost column we note the number of the move that is attacked. When writing the moves we use some abbreviations. When a formula is asserted, instead of writing "P-!-p∨¬p" for example, we simply write the formula "p∨¬p".

In the first row the Proponent is asserting the thesis p∨¬p, which counts as move number 0. In the first move the Opponent is attacking the disjunction according to the rule **PR∨**. According to this rule, it is the Proponent who has a choice of which disjunct he wants to defend, so he chooses ¬p in move 2. In move 3, the Opponent attacks the negation by asserting the non-negated formula p, according to the rule **PR¬**. The rule for negation states that there is no defence available to the Proponent, so the next row in his column is empty. However, the structural rule **SR 1.C** allows the Proponent to repeat his earlier defence against an attack even if it is not the last non-answered attack. Therefore, in move 4 the Proponent responds once more to the Opponent's attack against the disjunction from move 1, but this time he chooses the left disjunct p. The Opponent is not really repeating his attack, so we note it with brackets, just to make it clear to what move this defence corresponds. The atomic formula p is now repeated both by the Opponent and the Proponent, so that the Proponent wins the dialogue. The law of the excluded middle holds for classical first-order logic.

Let us now observe an example of an intuitionistic dialogue for the same formula:

	Opponent			Proponent	
				p∨¬p	0
1	?-∨		0	¬p	2
3	p		2		

As before, the Proponent responds to the Opponent's attack on the disjunction by choosing the right disjunct. In the following step, he does not have a proper defence against the Opponent's attack against negation. Only this time the dialog ends there. According to the structural rule **SR 1.I**, the Proponent does not have the right to repeat the defence against the previous attack by the Opponent. Remember that, according to this rule, only the last attack that has still not been answered can be defended. This time the dialogue is open and the Proponent loses the game. The law of the excluded middle is not valid in intuitionistic logic.

Bibliography

T. Bays (2014). "Skolem's Paradox", in Edward N. Zalta (ed.); *The Stanford Encyclopedia of Philosophy* (Winter 2014 Edition), URL = <http://plato.stanford.edu/archives/win2014/entries/paradox-skolem/>.

Jc. Beall, M. Glanzberg (2014). "Liar Paradox", in Edward N. Zalta (ed.); *The Stanford Encyclopedia of Philosophy* (Spring 2014 Edition), URL = <http://plato.stanford.edu/archives/fall2014/entries/liar-paradox/>.

J. Bell (2009). *The Axiom of Choice*. London: College Publications, 2009.

E. Bishop (1967). *Foundations of constructive mathematics*. New York, Toronto, London: McGraw- Hill, 1967.

A. Blass, Y. Gurevich (1986). "Henkin quantifiers and complete problems". *Annals of pure and applied logic*, 1986, vol. 32, pp. 1-16.

A. Blass (1992). "A game semantics for linear logic". *Annals of Pure and Applied Logic*, 1992. vol. 56, pp.183–220.

R. Brandom (1998). *Making It Explicit. Reasoning, Representing and Discursive Commitement.* Harvard University Press, 1998.

R. Brandom (2001). *Articulating Reasons. An Introduction to Inferentialism.* Harvard University Press, 2001.

X. Caicedo, F. Dechesne, T. M. V. Janssen (2009). "Equivalence and Quantifier Rules for Logic with Imperfect Information". *Logic Journal of the IGPL*, 2009, vol. 17(1), pp. 91–129.

P. Cameron, W. Hodges (2001). "Some Combinatorics of Imperfect Information". *Journal of Symbolic Logic*, 2001, vol. 66, pp. 673–684.

N. Clerbout, M.H. Gorisse, S. Rahman (2011). "Context-sensitivity in Jain philosophy: A dialogical study of Siddharsiganis *Commentary on the Handbook of Logic*". *Journal of Philosophical Logic*, 2011, vol. 40(5), pp. 633–662.

N. Clerbout, S. Rahman (2013). "On Dialogues, Predication and Elementary Sentences". *Revista de Humanidades de Valparaiso*, 2013, vol. 2, pp. 7 – 47.

N. Clerbout (2013a). "First-order dialogical games and tableaux". *Journal of Philosophical Logic*, 2013, doi: 10.1007/s10992-013-9289-z. URL http://dx.doi.org/10.1007/s10992-013-9289-z, pp. 1–17.

N. Clerbout (2013b). *Etude sur quelques sémantiques dialogiques: Concepts fondamentaux et éléments de metatheorie*. PhD thesis, Lille 3 / Leiden: Universities of Lille and Leiden, 2013.

N. Clerbout, S. Rahman (2014). *Linking Game-Theoretical Approaches with Constructive Type Theory: Dialogical Strategies as CTT-Demonstrations*. to appear.

P. J. Cohen (1963). "The Independence of the Continuum Hypothesis I". *Proceedings of the U.S. National Academy of Science,* vol 50, pp. 1143- 48.

P. J.Cohen (1964). "The Independence of the Continuum Hypothesis II". *Proceedings of the U.S. National Academy of Science,* vol 51, pp. 105-110.

R. Cook, S. Shapiro (1998). "Hintikka's Revolution: The Principles of Mathematics Revisited". *The British Journal for the Philosophy of Science*, 1998, vol. 49 (2), pp. 309 – 316.

Th. Coquand (2014). "Recursive functions and Constructive Mathematics", in J. Dubucs, M. Bourdeau (eds.); *Constructivity and Computability in Historical and Philosophical Perspective*. Dordrech: Springer, 2014.

F. Dechesne (2005). *Game, Set, Maths: formal investigations into logic with imperfect information*. PhD thesis, Tilburg: Tilburg University, 2005.

R. Dedekind (1996). "What are Numbers and What should they Be? ". *Journal of Symbolic Logic*, 1996, Vol. 61 (2), pp. 688-689.

R. Diaconescu (1975). "Axiom of choice and complementation". *Proc. Amer. Math. Soc.*, 1975, vol. 51, pp.176–178.

M. Ekland, D. Kolak (2002). "Is Hintikka's Logic First Order?". *Synthese*, 2002, vol. 131 (3), pp. 371 - 388.

H. Enderton (1970). "Finite partially ordered quantifiers". *Zeitschrift für mathematische Logik und Grundlagen der Mathematik*, 1970, vol. 16, pp. 393-397.

S. Feferman (2006). "What kind of logic is 'Independence Friendly' logic?", in R. E. Auxier, L. E. Hahn (eds.); *The Philosophy of Jaakko Hintikka.* Library of Living Philosophers, Open Court, 2006, vol. 30, pp. 453-469.

W. Felscher (1985). "Dialogues as a foundation for intuitionistic logic" in D. Gabbay and F. Guenthner (eds); *Handbook of Philosophical Logic*. Dordrecht: Kluwer, 1985, vol. 3, pp. 341–372.

W. Felscher (1994). "Review of Jean E. Rubin 'Mathematical logic: applications and theory'". *The Journal of Symbolic Logic*, 1994, vol. 59, pp. 670-671. doi:10.2307/2275418.

V. Fiutek, H. Rücket, S. Rahman (2010). "A dialogical semantics for Bonannos system of belief revision", in P. Bour and al. (eds); *Constructions*. London: College Publications, 2010, pp. 315–334.

M. Fontaine (2013). *Argumentation et engagement ontologique. Etre, c'est être choisi*. London: College Publications, 2013.

D. Gale, F. Stewart (1953). "Infinite games with perfect information". *Ann. of Mathematical Studies*, 1953, vol.28, pp. 245 – 266.

N. Gierasimczuk, J.Szymanik (2009). "Branching quantification v. two-way quantification". *Journal of Semantics*, 2009, vol. 26 (4), pp. 367 – 392.

J. Ginzburg (2012). *The Interactive Stance: Meaning for Conversation*. Oxford: Oxford University Press, 2012.

J.Y. Girard (1999). "On the meaning of logical rules I : syntax vs. semantics", in U. Berger, H. Schwichtenberg (eds); *Computational Logic*. Verlag, Heidelberg: Springer, 1999, pp. 215–272.

N. D. Goodman, J. Myhill (1978). "Choice implies excluded middle". *Zeitschrift für mathematische Logik und Grundlagen der Mathematik*, 1978, vol. 24.

K. Gödel (1939). "Consistency proof for the generalized continuum hypothesis", *Proceedings of the National Academy of Sciences, U.S.A.*, vol. 25, pp. 220–224. Reprinted in Gödel (1990), pp. 28–32.

K. Gödel (1990). *Collected Works. II: Publications 1938–1974*. S. Feferman, J. Dawson, S. Kleene, G. Moore, R. Solovay, and J. van Heijenoort (eds.); Oxford: Oxford University Press, 1990.

J. Granström (2011). *Treatise on Intuitionistic Type Theory*. Dordrecht: Springer, 2011.

M. Hand (1993). "A defence of Branching Quantification". *Synthese*, 1993, vol. 95, pp. 419 – 432.

G. Heinzmann (2004). "Some Coloured Remarks on the Foundations of Mathematics in the 20th Century". *Logic, Epistemology and The Unity of Science*, 2004, vol. 1, pp. 41-50.

G. Heinzmann (2004). "Comments on Jaakko Hintikka's Post-Tarskian Truth". *Alternative Logics. Do sciences need them?*, 2004, pp. 165-173.

A. Heyting (1962). "After Thirty Years". *Logic, methodology and philosophy of science*. Stanford: Stanford University Press, 1962, vol. 289, pp. 194–197.

L. Henkin (1961). "Some remarks on infinitely long formulas". *Infinitistic Methods, Proceedings of the Symposium on Foundations of Mathematics, Warsaw, 2-9 September, 1959*, 1961, pp. 167–183.

J. Hintikka (1962). *Knowledge and Belief*. Ithaca, N.Y.,Cornell University Press.

J. Hintikka (1968). "Language-Games for Quantifiers". *American Philosophical Quarterly Monograph Series 2: Studies in Logical Theory*. Oxford: Basil Blackwell, pp. 46–72.

J. Hintikka (1973). *Logic, Language-Games and Information: Kantian Themes in the Philosophy of Logic*. Oxford: Clarendon Press, 1973.

J. Hintikka (1979). "Language Games", in Saarinen (ed.); *Game Theoretical Semantics*. Dordrecht: D. Reidel, 1979, pp. 1 – 26.

J. Hintikka, J. Kulas (1985). *Anaphora and Definite Descriptions, Two Applications of Game- Theoretical Semantics*. Dordrecht, Boston, Lancaster: D. Reidel publishing company, 1985.

J. Hintikka, G. Sandu (1989). "Informational Independence as a Semantical Phenomenon", in J. E. Fenstad, I. T. Frolov, R. Hilpinen (eds.); *Logic, Methodology and Philosophy of Science*. Amsterdam: Elsevier, 1989, vol. 8, pp. 571–589.

J. Hintikka (1991). *Defining Truth, the Whole Truth and Nothing but the Truth*, Reports from the Department of Philosophy, No. 2, Helsinki: University of Helsinki, 1991.

J. Hintikka (1996a). *The Principles of Mathematics Revisited*. Cambridge University Press, 1996.

J. Hintikka (1996b). *Lingua Universalis vs. Calculus Ratiocinator: An Ultimate Presupposition of Twentieth-Century Philosophy*. Dordrecht: Kluwer, 1996.

J. Hintikka (1997a) "No Scope for Scope? ". *Linguistics and Philosophy*, 1997, vol. 20, pp. 515–544.

J.Hintikka (1997b). "Hilbert Vindicated?". *Synthese*, 1997, vol. 110, No. 1, pp. 15-36.

J. Hintikka, G. Sandu (1997). "Game-theoretical semantics", in J. van Benthem and A. ter Meulen (eds.); *Handbook of Logic and Language*. Amsterdam: Elsevier, 1997. pp. 361–410.

J. Hintikka (1999). *Inquiry as Inquiry: A Logic of Scientific Discovery*. Dordrecht: Springer, 1999.

J. Hintikka, I. Halonen, A. Mutanen (1999). "Interrogative Logic as a General Theory of Reasoning", in J. Hintikka, *Inquiry as Inquiry: A Logic of Scientific Discovery*. Dordrecht: Springer, 1999, pp. 47–90.

J. Hintikka (2000). "Game Theoretical Semantics as a Challenge to Proof Theory". *Nordic Journal of Philosophical logic,* vol. 4, No 2, pp. 127 – 141.

J. Hintikka (2001). "Intuitionistic Logic as Epistemic Logic". *Synthese*, 2001, vol. 127 (1-2), pp. 7 – 19.

J. Hintikka (2002a). "Hyperclassical Logic (A.K.A. IF Logic) and Its Implications for Logical Theory". *The Bulletin of Symbolic Logic*, 2002, vol. 8, No. 3, pp. 404-423.

J. Hintikka (2002b). "Negation in logic and in natural language". *Linguistics and Philosophy*, 2002, vol. 25, pp. 585 – 600.

J. Hintikka (2006a). "Reply to Wilfrid Hodges", in R. E. Auxier, L. E. Hahn (eds.); *The Philosophy of Jaakko Hintikka*. Library of Living Philosophers, Open Court, 2006, vol. 30, pp. 535 – 540.

J. Hintikka (2006b). "A Proof of Nominalism: An Exsercise in Successful Reduction in Logic". *From ontos verlag*: *Publications of the Austrian Ludwig Wittgenstein Society*, New Series, 2006, vol. 1.

J. Hintikka (2006c). "Truth, Negation and Other Basic Notions of Logic", in van Benthem, Heinzman, M. Rebushi, H. Visser (eds.); *The Age of Alternative Logics.* Springer, 2006.

J.Hintikka(2008)."Reforming Logic (and Set Theory)". http://people.bu.edu/hintikka/Papers_files/Hintikka.Reforming%20logic.and%20se t%20theory.0408.pdf (8 july, 2014, 10:55 am)

J. Hintikka (2011). "What is the Axiomatic Method?". *Synthese*, 2011, vol. 183, pp. 69 – 85.

W. Hodges (1997a). "Compositional semantics for a language of imperfect information". *Logic Journal of the IGPL*, 1997, vol. 5, pp. 539–563.

W. Hodges (1997b). "Some strange quantifiers", in J. Mycielski, G. Rozenberg, A. Salomaa, (eds); *Structures in Logic and Computer Science: A Selection of Essays in Honor of A. Ehrenfeucht.* Lecture Notes in Computer Science, Verlag: Springer, 1997, pp 51–65.

W. Hodges (2001). "Dialogue foundations. a sceptical look". *Proceedings of the Aristotelian Society*, 2001, vol. LXXV, pp. 17–32.

W. Hodges (2013). "Logic and Games", in Edward N. Zalta (ed.); *The Stanford Encyclopedia of Philosophy.* (Spring 2013 Edition), URL = <http://plato.stanford.edu/archives/spr2013/entries/logic-games/>.

T. Janssen (2002). "Independent choices and the interpretation of IF-logic". *Journal of Logic, Language and Information*, 2002, vol. 11, pp. 367–387.

T. Jansenn, F.Dechesne (2006). "Signalling in IF Games: A Tricky Business". *The Age of Alternative Logics,* 2006, part IV, pp. 221-241.

R. Jovanović (2013). "Hintikka's Take on the Axiom of Choice and the Constructivist Challenge". *Revista de Humanidades de Valparaiso*, 2013. No 2, pp. 135 – 152.

W. Kamlah, P. Lorenzen (1972). *Logische Propüadeutik.* Stuttgart, Weimar: Metzler, second edition, 1972.

W. Kamlah, P. Lorenzen (1984). *Logical Propaedeutic.* Lanham Md. University Press of America, English translation of Kamlah and Lorenzen (1972) by H. Robinson.

L. Keiff (2007). *Le Plralisme Dialogique: Approches dynamiques de largumentation formelle*. PhD thesis, Lille 3, 2007.

L. Keiff (2009). "Dialogical Logic", in Edward N. Zalta (ed.); *Stanford Encyclopedia of Philosophy* (Summer 2011 Edition), http://plato.stanford.edu/entries/logic-dialogical/.

S. Kleene (1967). *Mathematical Logic*. New York: John Wiley & Sons, 1967.

A. Lecomte, M. Quatrini (2010). "Pour une étude du langage via l'interaction: Dialogues et sémantique en Ludique". *Mathématiques et Sciences Humaines*, 2010, vol. 189, pp. 37–67.

A. Lecomte, M. Quatrini (2011). "Figures of dialogue: A view from Ludics". *Synthese*, 2011, vol.183(1), pp. 59–85.

A. Lecomte (2011). *Meaning, Logic and Ludics*. London: Imperial College Press, 2011.

A. Lecomte, S. Troncon (eds); (2011). *Ludics, Dialogues and Interaction: PRELUDE Project 2006–2009. Revised Selected Papers*. Berlin, Heidelberg: Springer, 2011.

K. Lorenz (1970). *Elemente der Sprachkritik Eine Alternative zum Dogmatismus und Skeptizismus in der Analytischen Philosophie*. Frankfurt:Suhrkamp Verlag, 1970.

K. Lorenz (2001). "Basic objectives of dialogue logic in historical perspective". *Synthese*, 2011, vol. 127 (1-2), pp. 255–263.

K. Lorenz (2010a). *Logic, Language and Method: On Polarities in Human Experience*. Berlin, New York: De Gruyter, 2010.

K. Lorenz (2010b). *Philosophische Variationen: Gesammelte Aufstze unter Einschluss gemeinsam mit Jürgen Mittelstraß geschriebener Arbeiten zu Platon und Leibni*. Berlin, New York: De Gruyter, 2010.

P. Lorenzen, O. Schwemmer (1975). *Konstruktive Logik, Ethik und Wissenschaftstheorie*. Mannheim: Bibliographisches Institut, second edition, 1975.

P. Lorenzen, K. Lorenz (1978). *Dialogische Logik*. Darmstadt: Wissenschaftliche Buchgesellschaft, 1978.

P. Lorenzen (1995). *Einführung in die operative Logik und Mathematik*. Berlin: Springer, 1995.

S. Magnier (2013). *Approche dialogique de la dynamique épistémique et de la condition juridique*. London: College Publications, 2013.

A. Mann, G. Sandu, M. Sevenster (2011). *Independence-Friendly Logic: A Game-Theoretic Approach*. London Mathematical Society Lecture Note Series: 386, 2011.

M. Marion (2006). "Hintikka on Wittgenstein : From Language-Games to Game Semantics", in T. Aho, A.V. Pietarinen (eds); *Truth and Games, Essays in Honour of Gabriel Sandu*. Acta Philosophica Fennica, 2006, vol. 78, pp. 223-242.

M. Marion (2009). "Why Play Logical Games ?", O. Majer, A.V. Pietarinen, T. Tulenheimo (eds.); *Logic and Games, Foundational Perspectives*. Dordrecht: Springer, 2009, pp. 3-26.

P. Martin-Löf (1984). *Intuitionistic Type Theory. Notes by Giovanni Sambin of a series of lectures given in Padua, June 1980*. Naples: Bibliopolis, 1984.

P. Martin- Löf (2006). "100 years of Zermelo's axiom of choice: what was the problem with it?". *The Computer Journal*, 2006, vol 49 (3), pp. 345–350.

D. McAdams, J. Sterling (2015). arXiv: 1410.4639 [cs.CL].

C. McCarty, N. Tennant (1987). "Skolem's Paradox and Constructivism". *Journal of Philosophical Logic*, 1987, vol. 16, pp. 165 – 202.

J. von Neumann, O. Morgenstern (1944). *The Theory of Games and Economic Behavior*. Princeton: Princeton University Press, 1944.

M. Pantasar (2009). *Truth, proof and G.de lian arguments : a defence of Tarskian truth in mathematics*. PHD thesis, Helsinki: University of Helsinki, Departement of philosophy, 2009.

T.E.Patton (1991). "On the Ontology of Branching Quantifiers". *Journal of Philosophical Logic*, 1991, vol. 20, pp. 205 – 223.

C. S. Peirce (1898). *Reasoning and the Logic of Things: The Cambridge Conferences Lectures of 1898*, in Kenneth Laine Ketner (ed.); Cambridge Mass: Harvard University Press, 1992.

A. Popek (2012). "Logical dialogues from Middle Ages", in C. Barés and al. (eds);

Logic of Knowledge. Theory and Applications. London: College Publications, 2012, pp. 223–244.

D. Prawitz (1979)."Proofs and the meaning and completeness of the logical constants", in J. Hintikka et al. (eds.); *Essays on Mathematical and Philosophical Logic*. Dordrecht: Reidel, 1979, pp. 25–40.

D. Prawitz (2012). "Truth and Proof in Intuitionism", in P. Dybjer, S. Lindström, G. Sundholm, E. Palmgren (eds.); *Epistemology versus Ontology: Essays on the Philosophy and Foundations of Mathematics in Honour of Per Martin-Löf.* Dordrecht: Springer, 2012, pp. 45-68.

G. Primiero (2008). "Constructive modalities for information", talk given at the*Young Researchers Days in Logic, Philosophy and History of Science.* Brussels,1-2 September 2008.

G. Primiero (2012). "A contextual type theory with judgmental modalities for reasoning from open assumptions". *Logique et Analyse*, 2012, vol. 220, pp. 579–600.

S. Rahman (1993). *Uber Dialoge, Protologische Kategorien und andere Seltenheiten.* Frankfurt, Paris , New York: P. Lang, 1993.

S. Rahman, L. Keiff (2005). "On how to be a dialogician", in D. Vanderveken (ed.); *Logic, Thought, and Action*. Dordrecht: Kluwer, 2005, pp. 359–408.

S. Rahman, T. Tulenheimo (2006). "From Games to Dialogues and Back: Towards a General Frame for Validity", in O. Majer and al. (eds); *Games: Unifying Logic,Language and Philosophy*. Dordrecht: Springer, 2006, pp. 153–208.

S. Rahman, L. Keiff (2010). "La Dialectique entre logique et retorique". *Revue de Métaphysique et de Morale*, 2010, vol. 66(2), pp. 149–178.

S. Rahman (2012). "Negation in the Logic of first degree entailment and *tonk*: A dialogical study", in S. Rahman, G. Primiero, M. Marion (eds); *The RealismAntirealism Debate in the Age of Alternative Logics*. Dordrecht: Springer, 2012, pp. 213–250.

S. Rahman, N. Clerbout (2013). "Constructive Type Theory and the Dialogical Approach to Meaning". *The Baltic International Yearbook of Cognition, Logic and Communication,* 2013, vol. 8: Games, Game Theory and Game Semantics, pp. 1-72. DOI: 10.4148/1944-3676.1077. Also online in: www.thebalticyearbook.org.

S. Rahman, N. Clerbout (2014). *Linking Games and Constructive Type Theory: Dialogical Strategies, CTT-Demonstrations and the Axiom of choice.* Springer, forthcomming.

S. Rahman, N. Clerbout, Z. McConaughey (2014). "On play-objects in dialogical games. Towards a Dialogical approach to Constructive Type Theory", in P. Allo (ed.); *Tribute to J.P. van Bendegem*. College Publications, to appear.

S. Rahman, N. Clerbout, R. Jovanović (forthcoming). "Knowledge and its Game Theoretical Foundations: The Challenges of the Dialogical Approach to Constructive Type Theory". Springer, forthcomming.

A. Ranta (1988). "Propositions as games as types" *Synthese*, 1988, vol. 76, pp. 377–395.

A. Ranta (1991). "Constructing possible worlds". *Theoria*, 1991, vol. 57(1-2), pp. 77–99.

A. Ranta (1994). *Type-Theoretical Grammar*. Oxford: Clarendon Press, 1994.

S. Read (2008). "Harmony and modality", in C. Dégremont et al. (eds.); *Dialogues, Logics and Other Strange Things: Essays in Honour of Shahid Rahman*. London: College Publications, 2008, pp. 285–303.

S. Read (2010). "General Elimination Harmony and the Meaning of the Logical Constants". *Journal of Philosophical Logic*, 2010, vol. 39 (5), pp. 557–576.

M. Rebushi (2010). "Heinzmann, Hintikka, et la vérité", in P.E. Bour, M. Rebuschi and L. Rollet (eds.) ; *Construction. Festschrift for Gerhard Heinzmann*. London, College Publications, pp. 387-396.

J. Redmond (2010). *Logique dynamique de la fiction: Pour une approche dialogique*. London:College Publications, 2010.

J. Redmond, M. Fontaine (2011). *How to Play Dialogues: An Introduction to Dialogical Logic*. London: College Publications, 2011.

M. Resnik (1966). "On Skolem's Paradox". *The Journal of Philosophy*, 1966, vol. 63, pp. 425–438.

H. Rückert (2011a). *Dialogues as a Dynamic Framework for Logic*. London: College Publications, 2011.

H. Rückert (2011b). "The Conception of Validity in Dialogical Logic", Talk at the workshop *Proofs and Dialogues*. Tübingen, 2011.

P. de Rouilhan, S. Bozon (2006). "The Truth of IF: Has Hintikka Really Exorcised Tarski's Curse? ". Auxier & Hahn, 2006, pp. 683–705.

H. Ruckert (2011). *Dialogues as a Dynamic Framework for Logic.* Dialogues and Games of logic, vol 2, College Publications, ISBN 1848900473, 9781848900479, 2011.

E. Saarinen (1978). "Dialogue semantics versus game-theoretical semantics". *Proceedings of the Biennial Meeting of the Philosophy of Science Association* Vol. 2: *Symposia and Invited Papers.* Chicago: University of Chicago Press, 1978, pp. 41–59.

G. Sandu (1991). *Studies in Game-Theoretical Logics and Semantics.* Ph.D. thesis, Helsinki: University of Helsinki, 1991.

G. Sandu (1997). "On the theory of anaphora: dynamic predicate logic vs. game-theoretical semantics". *Linguistics and Philosophy*, 1997, vol. 20, pp.147-174.

G. Sandu, J. Jacot (2012). "Quantification and Anaphora in Natural Language". *Prospect for meaning*, Berlin, New York: Walter de Gruyter Inc, 2012.

P. Schroeder-Heister (2008). "Lorenzen's operative justification of intuitionistic logic", in M. van Atten et al. (eds.); *One Hundred Years of Intuitionism (1907– 2007).* Basel: Birkhäuser, 2008, pp. 214–240.

S. Shapiro (1991). *Foundations without Foundationalism.* Oxford: Oxford University Press, 1991.

T. Skolem (1955). "A critical remark on foundational research". *Norske Vid. Selsk. Forh., Trondheim,* 1955, vol 28, pp.100–105.

G. Sundholm (1986). "Proof-theory and meaning", in D. Gabbay, F. Guenthner (eds.); *Handbook of Philosophical Logic*, vol. 3, Dordrecht: Reidel, 1986, pp. 471– 506.

G. Sundholm (1997). "Implicit epistemic aspects of constructive logic". *Journal of Logic, Language, and Information,* 1997, vol. 6 (2), pp. 191-212.

G. Sundholm (1998). "Inference versus Consequence", in T. Childers (ed.), *The Logica Yearbook 1997*,Prague:Filosofia, pp. 26–36.

G. Sundholm (2001). "A Plea for Logical Atavism", in O. Majer (ed.); *The Logica Yearbook 2000.* Prague: Filosofia, 2001, pp. 151-162.

G. Sundholm (2009). "A Century of Judgment and Inference: 1837–1936", in

L. Haaparanta (ed.); *The Development of Modern Logic*. Oxford: Oxford University Press, 2009, pp. 263–317.

G. Sundholm (2013a). "Independence Friendly Language is First - order after all?". *Logique et Analyse*, 2013.

G. Sundholm (2013b). "Inference and Consequence in an Interpeted Language". Talk at the Workshop *Proof theory and Consequence in an Interpreted Language.* Groningen, December 3-5, 2013.

G. Sundholm (2014) "Constructive Recursive Functions, Church's Thesis and Brower's Theory of the Creating Subject: Afterthoughts on a Parisian Joint Session", in J. Dubusc and M. Bourdeau (eds.); *Constructivity and Computability in Historical and Philosophical Perspective*. Logic, Epistemology and The Unity of Science, vol. 34, Dordrecht: Springer, 2014.

W. Tait (1994). "The law of excluded middle and the axiom of choice", in A. George (ed.); *Mathematics and Mind*, New York: Oxford University Press, 1994, pp. 45–70.

A. Tarski (1935). translation by J.H. Woodger (1983). "The Concept of Truth in Formalized Languages", in J. Corcoran (ed.); *Logic, Semantics, Metamathematics*. Hackett, 1935.

N. Tennant (1998). "Game Some People Would Have All of Us Play". *Philosophia Mathematica*, 1998, vol. 6 (3), pp. 90 - 115.

T. Tulenheimo (2009). "Independence Friendly Logic", in Edward N. Zalta (ed.); *The Stanford Encyclopedia of Philosophy (Summer 2009 Edition)*, URL = <http://plato.stanford.edu/archives/sum2009/entries/logic-if/>.

T. Tulenheimo (2011). "On Some Logic Games in Their Philosophical Context", in A. Lecomte, S. Tronçon (eds.); *Ludics, Dialogues and Interaction: PRELUDE Project 2006–2009, Revised Selected Papers*. Berlin, Heidelberg: Springer, 2011, pp. 88–113.

M. Van Atten (2014). "Gödel and Intuitionism", in J. Dubusc and M. Bourdeau (eds.); *Constructivity and Computability in Historical and Philosophical Perspective*. Logic, Epistemology and The Unity of Science, vol. 34, Dordrecht: Springer, 2014.

J. Van Benthem (1996). *Exploring Logical Dynamics*. Stanford Calif.,Cambridge: CSLI Publications and Cambridge University Press, 1996.

J. Van Benthem (2001a). "Correspondence Theory", in D. Gabbay, F. Guenthner (eds.); *Handbook of Philosophical Logic*. Volume 2, Dordrecht: Reidel, 2001, pp. 167–247, Reprint with addenda in second edition, pp. 325–408.

J. Van Benthem (2001b) "Logic in Games". Lecture notes, unpublished.

J. Van Benthem(2011). *Logical Dynamics of Information and Interaction*. Cambridge: Cambridge University Press, 2011.

H. Van Ditmarsch, W. van der Hoek, B. Kooi (2007). *Dynamic Epistemic Logic*. Berlin: Springer, 2011.

J. Väänänen (2001). "Second-Order Logic and Foundations of Mathematics". *Bulletin of Symbolic Logic*, 2001, vol. 7, pp. 504–520.

W. Walkoe (1970). "Finite partially ordered quantification". *Journal of symbolic logic*, 1970, vol. 35, pp. 535-555.

E. Zermelo (1908). "Untersuchungen über die Grundlagen der Mengenlehre, I". *Mathematische Annalen*, 1908, vol. 65, pp. 261 – 281.

Index of names and notions

www.ingramcontent.com/pod-product-compliance
Lightning Source LLC
LaVergne TN
LVHW012330060326
832902LV00011B/1804